Electronic Security Systems

Electronic Security Systems

Reducing false alarms

Philip Walker BSc(Eng), MIEE

Third Edition

Newnes

OXFORD BOSTON JOHANNESBURG MELBOURNE NEW DELHI SINGAPORE

Newnes
An imprint of Butterworth-Heinemann
Linacre House, Jordan Hill, Oxford OX2 8DP
225 Wildwood Avenue, Woburn, MA 01801–2041
A division of Reed Educational and Professional Publishing Ltd

 A member of the Reed Elsevier plc group

First published 1983
Second edition 1988
Third edition 1998

Transferred to digital printing, 2005

© P. H. Walker 1998

British Library Cataloguing in Publication Data
A catalogue record for this book is available from the British Library.

ISBN 07506 3543 6

Library of Congress Cataloguing in Publication Data
A catalogue record for this book is available from the Library of Congress.

FOR EVERY VOLUME THAT WE PUBLISH, BUTTERWORTH-HEINEMANN
WILL PAY FOR BTCV TO PLANT AND CARE FOR A TREE.

Composition by Scribe Design, Gillingham, Kent
Printed & bound by Antony Rowe Ltd, Eastbourne

Contents

Foreword

This book is being published at a time when governments, commerce and industry, and indeed ordinary householders, have to pay more attention to security than they have ever done before, because the crime epidemic is virtually out of control.

For some time now police chiefs everywhere have been making it abundantly clear, in public, that the police alone cannot cope with the problem. Forcible entry to premises has become a boom industry, and visits to the scenes of crime by police are no more than perfunctory, even at a time when police recruitment has been very substantially increased and the technical aids available to assist them in their task are better than ever before. It seems their best endeavours are doomed to failure unless every single one of us takes steps now to protect ourselves and our assets.

Insurance companies are now refusing to cover risks in some city areas. In purely financial terms, the insurance industry is only marginally concerned with crime prevention and only a small proportion of its total income is derived from insuring burglary and theft risks, and perhaps insurance companies also feel they are fighting a losing battle in the industrial and domestic sectors, although for many years they have been constantly cajoling policy holders to improve security standards.

Where do we go from here?

It is now more apparent than ever before that the increasing complexity of commerce and industry means that there are ever-increasing risks to be combated, and this requires that security become much more professional and effective: the problem will only be resolved when all those agencies directly involved in it unite and become fully acquainted with the major contribution that electronic security systems, correctly installed in the right environments, will make to the overall security function.

I think that the requirements of customers have for far too long been overlooked, maybe because they, in the main, know very little about security techniques. This particularly applies to technical matters, and consequently management delegate the task to whomever professes to know something about it, or promises to find out: the consequences are now with us.

The problem hitherto has been the absence of a really comprehensive manual analysing risks, equipment, system design, and user requirements available for quick and easy reference, which poses the problems and offers the solutions in a manner which the layman will also understand and appreciate.

We now have such a book. The author, Philip Walker, clearly has an abundant expert knowledge of electronic security systems, and, being uniquely qualified both academically and in practice, realises that the vast majority of people know very little about electronic security systems, how to install them effectively and how they detect intruders. More important, perhaps, he possesses a flair for communicating the complex and intricate functions performed by security devices in a manner capable of being understood by ordinary people involved in security, and because of this I believe the book will eventually become widely read. The result is an invaluable text which will appeal to a very large readership, including customers, installers, insurance surveyors, security officers, police crime prevention officers and, indeed, anyone who is in any way connected with security.

The information imparted could indirectly save commerce and industry everywhere vast amounts of money, and prevent disruption of business and, perhaps, loss of customers, when it is fully realised that it is possible to install security systems which can be stable and effective in order to prevent crime.

The insurance industry should welcome the book's publication because surveyors armed with the wealth of knowledge contained therein will be able to insist on more suitable protection for the risks they are insuring, and do so with confidence.

In conclusion, and perhaps most importantly, the police should applaud its arrival at a time when crime is at an all-time high and they are sadly disillusioned after so many years attending alarm signals from electronic security systems only to find that even now, nearly ninety times out of a hundred their time has been totally wasted, because of false alarms being triggered through equipment installed in the wrong environment, equipment failure, poor alarm system design or human error and negligence. If Philip Walker's advice could assist in eliminating this drain on police resources alone, it would have achieved a great deal.

Alec Thomson MBE
Security Adviser

Preface

Well, well, here we are again. What has been happening in the security world in the years between the publication of the second and third editions? For one thing, the book was sold out, and the publishers, Butterworth-Heinemann, were being asked for more.

In my so called retirement I had started a small company, CPA Systems, designing and supplying portable public address systems for National Trust local lectures and the like. After Sue Brown had coxed a Boat Race crew to victory, rowing became a new sport for girls. An American firm had pioneered the supply of voice amplification systems, stroke rate meters etc. for rowing, so we decided to give them a bit of competition, and to give the girls something better. So when Butterworth-Heinemann started saying they wanted a third edition, my answer must have been on the lines of 'No way – I'm busy.' But they persevered, and I suggested they used a ghost writer – I would feed him and he could do the donkey work. Not surprisingly really, there were no takers.

Back came Butterworth-Heinemann, this time with the suggestion that the third edition be published through their associated company, Newnes. I had always hankered after this; Butterworth-Heinemann are famous for academic and scientific books, while Newnes have a more practical-user slant – always my target readership. And so, I signed up to write the third edition.

Readers of previous editions will know that, in one way and another, communications and applied electronics have been my life, first in broadcasting with the BBC and later in radar with Decca. While with Decca I was asked whether my work on military radar could be applied to prison security, and gradually I moved into full-time work on security problems. With successive organisations I was involved with the design, manufacture, marketing, installation and use of electronic equipment for security. This perhaps unusual combination of experience led to work as a security consulting engineer, and to an invitation to write this book.

History doesn't change, but it needs to be brought up-to-date. So what else has happened in the security world? With the change of emphasis from the industrial to the domestic market the number of installation suppliers has mushroomed, but the equipment manufacturers seem to have consolidated into fewer and larger companies. There has been little development of new techniques in intruder detection devices, but valuable improvements in reliability have been made.

The one outstanding positive step forward has been the publication of the police ACPO Policy Document, aimed at concentrating minds at last on false alarm reduction. Results are already positive, and given no relaxation, results can be dramatic.

A diversion of valuable talent comes from the torrent of new standards being issued. I have written a new Chapter (28) to indicate the nature and scale of the situation, as well as to guide readers in finding which standards refer to their work.

BSIA and NACOSS with BSI have produced a competent and workable set of standards and codes of practice. Unfortunately, skills badly needed for continued implementation of the ACPO policy are being diverted by yet further standards, being issued by the EC. Time is a healer, and it cannot be long now before real progress is evident towards eliminating false alarms, in allowing the police to concentrate on real alarms, and so help to fulfil our objective – crime prevention.

My grateful thanks go to the many individuals who have contributed in their way to the making of the book, and also to Susan Garratt, the daughter of a former Head of the Science Museum's Communication Collections, including the radio station, for her help in preparing the book for publication.

But probably the greatest contribution was from my wife, Christine, whose encouragement and forbearance made it all possible.

Philip Walker

Part 1

Systems

1 What this book is about and how to use it

The security environment

'Security precautions were very tight.' A report of that nature throws up a wide range of thoughts, questions and implications. If in fact it was a news item, the implication is either that a very exceptional risk was covered or perhaps that, even now, it is rare for proper security precautions to be taken.

For some ideas on answers, we have to look at the local, national and international scenes, and see whether the civil, political and military pressures for villainy are increasing and whether the forces for law and order are maintaining sufficient control. That fundamentally is the security environment in which we work.

You might say: 'It is not for me to think about things like that – I just have to do as I'm told'. To which I would say, 'So be it, but don't you think you would make an even better security man if you understood a little of what goes on outside your immediate duties?'

So awareness is one thing this book is about. Ostensibly it is about electronics, but it is also about people. Did you know, for instance, that people and awareness were considered so important by General Montgomery during World War II that he insisted upon war correspondents being given more information than they would be allowed to publish? By understanding, rather than guessing, what was going on, the correspondents were less likely inadvertently to breach security and to give away information helpful to the enemy.

The user as the focal point

A horror of mine is the technically perfect but operationally useless device. This more than anything has led to my favourite definition of an engineer: he who starts with the user and works inwards towards the source, be that the laboratory or the customer. The scientist, on the other hand, starts in the laboratory or source, and works outwards. The engineer and the scientist, hopefully, meet somewhere in the middle.

Who, then, is the user? Most security equipment exists to give a warning that conditions have changed from those required. To me, the person there to receive the warning is the user. But the warning does not stay with the user – he has to do something about it. Depending upon his operational instructions, the user may well be responsible for triggering off a pretty serious and extensive chain of events, and against the clock. The user inevitably becomes the focal point in the event of an incident, and as such in my view he, or his function, must be the starting point in any study of a security problem.

The first question in such a study is, 'What is to be done with a warning signal?' A controversial definition of management is 'they who clear the way for the operational people to do their job properly'. All too often, when management are asked what the user is to do with a warning signal, and how is he to do it, they have not thought through the procedure, or the consequences.

No wonder, then, that some clients have been surprised when I have concentrated on sorting out the user and his problems before even seeing the property at risk or the security problem itself.

This is perhaps sufficient at this stage to emphasise the need for working inwards from the user rather than outwards, and the theme will crop up again in later chapters, particularly Chapters 2, 3 and 8. We shall see that the user may be miles away, and may belong to an independent industrial, civil or military police force, but his significant role is unchanged, no matter where he is.

Electronics as an aid to the user

Yes: an aid. To get the feel of this concept, think of the policeman out on patrol. Where once he would have used a whistle to attract help when he needed it, he now uses radio. Or think of an airport, where an electronic metal detector helps to find unwanted weapons being taken on board. The list is growing all the time, but throughout the book the view is held that electronic equipment is there to help the security officer to get information more accurately and more quickly, and to enable him to pass on the information for corrective action. So long as villains are human beings, we shall need human beings somewhere in the chain to deal with them. And electronic aids are added to enable security officers to do their jobs better, but not to do their jobs *for* them.

Security and the customer

No matter how much importance we give to the user in the planning stages, he just will not get his electronic aids unless someone buys them for him. Another purpose of this book is to help the customer and his advisers to understand what is needed, and to help managers, designers, surveyors, salesmen, installers and maintainers to understand the use of electronics in security well enough to get and implement the order to supply the electronic aids, and to help them to satisfy the customer and his operational personnel.

General arrangement of the book

Have you ever tried to design a filing system? A sales office is so single-minded that probably filing by customer name is enough, but in a design office, for instance, does one file by name of supplier, or by product, or by function? There does not seem to be a satisfactory answer if ready access to information is to be achieved, except by using duplication or even triplication. The same dilemma applies in arranging the chapters of a book on security, because of the diverse interests of the wide range of types of people involved.

What I have done, therefore, is to look at the subject from three main points of view.

In Part 1, 'Systems', we look at risk analysis, of interest to the surveyor and the customer; and at ways of combating a range of different types of risk, of interest to the system designer, the customer again, and the security officer – the user.

In Part 2, 'Equipment', the pros, cons and application of a variety of products representing the armoury available to the system designer are examined. Here also we try to spot gaps in the armoury, and look for possible ways of filling the gaps, all of which should be of particular interest to the equipment designer and the systems designer.

In Part 3, 'Implementation', the circle is completed by dealing with installation, operation and maintenance, which bring in again the interests of the user and of the varied categories of field engineer, upon whom falls the main task of keeping the customer happy.

However, no one part of the book is aimed only at one category or job title. The more each individual understands the work and problems of his colleagues in other aspects of security, the better his own work will be. Each of the three parts of the book is therefore divided into numerous chapters, and each chapter is given as many paragraph headings as is practicable to make it easy for the reader to find his way around. The duplication aspect mentioned above is covered via the index, where a reference to any item may well be covered from different viewpoints in two or all three parts of the book.

There ought to be a label on top of this chapter saying 'Read this first'. If by chance you are doing so, then guidance on your choice of reading order may be helpful.

For a broad commentary and understanding there is no better way than to start at the beginning, and to read right through to the end, jotting down notes as you go of uncertainties to be cleared up, of disagreements, and for future reference.

For more detailed study, please refer first to the remaining paragraphs of this chapter, dealing with its use as a handbook, a training aid and a source for ideas. 'For ideas' is probably a better phrase here than 'of ideas', as a little disagreement with an author is a good way of stimulating creative thinking.

Use as a handbook

This really means using the book as a reference book. There are so many aspects to security that even if you are totally familiar with, say, access control, you may be committed to a survey, or to a meeting, or to writing up a system specification in which you also have to deal with detection of an intruder who may manage to penetrate to a vulnerable point. Are you sure that you know how to decide on whether, for instance, you ought to use infrared, microwave ultrasonic or contact detectors? I was in just such a situation shortly after drafting one of the chapters, and I found myself reading up the draft just before going out on the job, to help me to have the pros and cons at my fingertips when I needed them.

Information is presented through the book as straight description, check lists, questions and answers, case histories, narrative, and so on, depending upon the aspect of security being dealt with. Much of it has to be my opinion, but if you read up the relevant section of the book shortly before becoming involved, you will be in that much stronger a position, whether you agree with my assessment or not, to crystallise your own views and to have a clearer-headed discussion with the other people likely to be involved.

You may find it odd that there are no circuit diagrams in this book, only the occasional block diagram. But the book is about applied electronics and not about electronics itself. Not only that, the book is about principles, and the application of principles, rather than about techniques. Electronic techniques change so rapidly that it is better for product designers to use the book to help them sort out 'what' the product is to do, and to obtain contemporary detailed design information elsewhere for deciding 'how' to do it.

Use in formal training and as a teach yourself aid

Unlike some of the clearly defined subjects in formal training, such as physics and chemistry, the use of electronics in security tends to spread across physics, chemistry and a number of other subjects, and it is relatively new.

Where electronics in security is taught in class, it is treated quite properly, as a part only of a wider aspect of security. In planning a syllabus, or in planning the detailed content of a course to meet a syllabus requirement, the general arrangement of this book, including paragraph headings and the index, makes it relatively easy to earmark various sections for study in class or at home. The teacher and the course organiser should themselves study as much as is practicable of the material preceding and following the items earmarked, in order to get the topic into context and perspective. The index should be used also to help form the basis of a recommended further reading list.

I am not sure that it is really true to say 'No impression without expression', but for an inexact subject such as security it is, I think, essential in group training classes to devote a longer time to discussion than is customary with the more factual subjects. Particular attention is needed in the training of supervisory staff to the need to be alert to the risks of negligence and the risk of wrong interpretation of instructions by their personnel. The difficulty of supervision is emphasised when it is realised that practical learning and experience, as well as operational duty, are on site rather than at base where supervision should be easy.

Electronics is expanding so rapidly in all walks of life that as the use of electronics in security becomes a subject in its own right, more of the book might be incorporated into a formal course. In fact, it should be practicable to write various types of syllabus round the book, rather than taking extracts from the book to match an existing syllabus.

As explained in Chapter 2, the whole of security is a risk-taking exercise, and it is emphasis rather than principle that tends to date. By keeping to principles rather than to detailed implementation, the reader is free to apply the principles in the best way at the time, and the information in the book should remain up to date for much longer.

For use as a teach yourself aid, the book is aimed at the practising security man who already has some experience and who has got past the stage of 'not knowing what he doesn't know'. Its use as a reference book or handbook will be particularly apt, but for more general advancement it is suggested that the reader should first skip through the book, to get the feel of the treatment and to find the areas where it strikes a chord with his own experience. It is almost a bonus if some areas strike a discord, for this provokes thought and discussion later with colleagues.

Once the structure of the book starts to take shape, it is worth remembering that it is hard to improve security without improving discipline. So a little self-discipline is recommended, and that might, after one has skipped through and given the matter thought, take the form of preparing one's own reading list based on chapter headings and contents, on the index, and on paragraphs that appear to be relevant to one's work at the time.

What is not covered is detailed instruction on installation practice. This is better based on the employer's own codes of practice and on the early parts of the relevant craft courses run by technical training schools and colleges.

Using the index as a glossary

For reasons we can only guess, security terminology is in rather a mess. When people talk to one another, it helps if each word used means the same thing to each person. Sadly, this isn't always so, and any group of people working regularly together should take unusual care to agree on the meaning of the words they use and to identify key words to be used to refer to particular situations. Naturally, the problem is rather worse when strangers try to convey information to one another, as you will soon discover as you read this book.

I have done two things about this problem. First, there is a chapter devoted to communication (Chapter 29) and then in the Index the pages on which I have tried to give the meanings of the words I have used are printed in **bold** type. These definitions are purely aids to understanding words used in this book, but anyone is welcome to adopt them for use elsewhere if they feel the meaning given is right. The Index is more detailed than usual, and includes some duplication to avoid the annoyance of cross-referencing. Please make a habit if you can of using the Index, and turn up each reference given, because the item may be treated from a different viewpoint in different parts of the book.

Finding information and ideas

Being first with the information is a strong motive, so, to keep up to date in one's field, the best sources of information are periodicals devoted to general and specialised aspects of the subject. Oddly enough, being first to publish can mean being too early. Quite often a new concept is put forward before the market is ready for it, and before development and manufacturing techniques have been evolved for it. The concept then can only be used years later, when techniques have caught up. So it is that reading periodicals several years old can be a fruitful source of ideas.

When, for instance, I was working on ultrasonic detection methods, some quite old publications on how that winged oddity the bat found its way in the dark helped by a process of elimination towards concentration on the Doppler principle. The tricks that the various breeds of bat get up to still makes fascinating reading, and man-made radar still falls short of the remarkable combination of detection techniques one small bat can use in its nightly life.

Which all brings us back to the value of reading books. Scattered through this book, for example, is information based on personal experiences that have not been published before, and on the experiences of others that may have been published in one form or another some time ago but are topical because only more recently have the techniques become practicable and applicable to security problems.

To come right up to date on what can be done now, there is no substitute for seeing the real, three-dimensional thing, at exhibitions, in suppliers' showrooms and at private demonstrations. The chapters in Part 2 of the book set out the principles of various types of equipment in a way that should help appreciation of the features of equipment as and when it is seen and used.

Come to think of it, you will not really appreciate the features until you have made Chapter 2, 'Thinking security', part of your everyday life. You cannot be a successful security man without *that* asset, so it seems to be a good place to start.

2 Thinking security

If you were a villain

My wife was amused at my reaction to a comment she made the other day about a window in our house. After completion of some minor building alterations, she suggested that it would no longer be necessary to lock that window. My reply was: 'OK, I'll go round and see how I would get in now.'

That simple episode illustrates the first steps in any security situation – awareness that a risk exists and assessment of that risk. One of the most valuable ways of making this assessment is to consider what you would do if you were a villain.

From the assessment, decisions can be made on what to do to safeguard the situation. In this chapter we shall take a rather closer look at the need for awareness and its consequences, and the steps will be examined in progressively greater detail in later chapters.

Awareness of a risk

If we are to talk about the need for awareness, we have to be quite clear as to what we are to be aware of. In security jargon the word 'risk' is used to express the possibility of unwanted change from the normal wanted situation that may be caused by a villain.

There are so many examples of things to be aware of that it is perhaps more effective to consider a few cases of unawareness and neglect. What do you say to the person who dismisses the subject by asserting, for instance, that 'It'll never happen to me'? Others are quite good at having 'Confidential' typed on the top of important papers, and ensuring that the papers are kept in a combination-locked safe; but are they aware that some typewriters use the ribbon only once and that their confidential story is spelled out like a telegram on the ribbon, for retrieval at will by a competent industrial spy?

Sheer habit and familiarity are other sources of risk. Take, for instance, a man working in a high-risk enclosure, equipped with access control facilities. Attending tests on some new equipment in one such place, I heard a door access bell ring, and a few moments later I said to a colleague beside me who was new to security, 'Did you see that?' 'No, what do you mean?', he replied. Quite simply, the regular attendant had released the access door before checking who was outside and wanting access. It always has been a 'friend' before – why should it be a 'foe' now? – sheer habit.

If you are doing a survey of some premises, you may see a cupboard door and ask your escort, 'What is behind that door?' 'Oh – that's all right, we always keep it locked', he replies. Your awareness was enough to see the cupboard door and to prompt the question – let it be enough also to get him to show you the lock and the contents, and to explain their system of key security. The chances are that you are doing the survey because someone there in authority is none too happy about their locks and keys anyway.

School yourself then in awareness: you cannot be a successful security man without it.

The dual question

Just as awareness is crucial for the surveyor in his risk analysis and proposals, the 'dual question' is crucial for the security system designer and installation team. Don't worry yet if you don't know what this means; I've had to invent the phrase because I don't know any one word that expresses the idea properly. The problem is that in security we work in a hostile environment. Someone tries to do something and we try to stop him. He tries to overcome what we have done, and we have to try harder.

Unless you believe in gremlins, there is really nothing and no one trying to interfere with the satisfactory operation of your mowing machine or sewing machine. However, the most popular of the early sensors in intruder detection systems died the death because frivolous, mischievous and evil individuals found that chewing-gum was very handy for jamming the spring-loaded plunger of mechanical door switch contacts into the position indicating 'door closed', when in fact the door was open.

With that as an example, we can see what the dual question tries to ask. It is not enough to ask in a security environment, 'Will it work satisfactorily?' We have also to ask, 'Can it be made not to work?' The mechanical door contact worked beautifully, but it was just too easily put out of action.

And so it has to be with ideas, equipment and systems used for all aspects of security. If the villain can beat you to it, the chances are that he will. So, in all decisions on actions to be taken, the best safeguard is to ask, and to get answers to the dual question, 'Will it work? Can it be made not to work?'

One robust character I had the pleasure of training used to become furious with me because, as he said, I so often criticised his ideas and substituted my own. Then quite suddenly, overnight it seemed, he got the message. So totally had he absorbed the concept of the dual question that I became the target, and more often than not it was he who won, with an idea for a better way.

Psychology and the villain

Without a villain we have not got a security problem. Conversely, if we have got a security problem, there must be a villain somewhere and we must try to understand how he will act. So, what makes a villain tick?

How can we know? Several former convicts have achieved widespread publicity by telling their stories of their crimes, why they did them and how they overcame the security measures put in their way. But can we believe what they say? They have defied the rules of law and order in the past; does their declared intention of 'going straight' now include total respect for the truth? We don't know, and the problem has been big enough to justify the establishment of Chairs of Criminology at leading universities to study crime and the criminal. Fortunately for our purposes, there is no need to penetrate as deeply as that, although undoubtedly there is much we can learn from these academic studies.

Deterrence

To start us off in the right direction, we can ask a few questions of ourselves, for perhaps we are not all that innocent-minded. What stops us from smashing something up, or from taking something that belongs to someone else – his car, or his wife, or his life? Probably it is a combination of, on the one hand, hopefully, a sound moral upbringing, and, on the other, the consequences of being found out.

The Latin word *deterrere*, meaning 'to frighten', is the origin of the English word which we now use – 'deter'. It is so apt in expressing the feeling of fear of the consequences that it is used to cover anything in security from the domestic window lock we mentioned earlier to the international deterrent, the nuclear bomb. Experience indicates that the villain is not immune to fear as a deterrent.

In security system design, therefore, we take into account the psychology of the villain, ideally by persuading him that it is not worth his while to make the attempt, or that if he does try, he will fail or be caught in the act.

Where the 'cause' is great enough in the mind of the villain, including situations that go beyond personal gain towards subversion (the overthrow of a political regime for instance), allowance has to be made where practicable for the attack going ahead in spite of the deterrent measures taken.

The concepts of deterrence, detection and protection will be expanded in later chapters.

Psychology and the user

In this context we have to use the word 'user' to cover the meaning in its widest sense, from the customer through to the people who are employed in various ways to make the security system work.

The basic feeling in the customer is usually that he doesn't want the thing anyway – he regards it as a non-productive expense. If the fear of the consequences of not taking security measures is not sufficiently strong in the customer for him to act on his own initiative, the decisive prod, more likely than not, will come from his insurance company, who have their own way of working on the fear concept. The insurance company may say that without their recommended security measures they would not give insurance cover of the risks; or if they are already giving cover of an increasing risk, that the premium costs would have to go up substantially unless security was improved.

There are limits to this form of persuasion, because there are other insurance companies and the user can shop around, just as for his car insurance, to find the best cover at the lowest price and, in this case, with the lowest requirements for security measures.

However, once the decision is made and security measures have been provided, the user will do well to reread his insurance policy. Somewhere in it he will find the requirement that not only must security measures be provided but also they must be used. That at least makes sense, but it does bring with it the responsibility for setting up and maintaining an operational security organisation, and delegation of responsibility for day-to-day running to the ultimate users.

Ideally, those concerned with day-to-day operation of security systems should welcome them, for they are aids to their work. Too often, by failure to recognise the psychology of the user, the provision of security aids causes resentment. This feeling can develop from fear that the aids may ultimately take over the users' jobs, or from changes needed in their routine following increased discipline created by the aids, or from aggravation caused by faulty behaviour of the aids.

Whatever the cause of resentment, the feeling can generate a deliberate or latent decision to ignore the aids, with the inevitable consequences if implementation of the decision coincides with activity by the villain. If a loss follows and a claim is made, and if the insurance company can establish that the security aids were not in use, the claim may be disallowed.

It is imperative, therefore, that in thinking security the psychology of the user be given a high place in the decision-making process. This topic is given more specific attention in Part 3, on Implementation.

Limits of responsibility

Heavy going though the last section may have been, nothing in it should allow us to be depressed into thinking that we have to bear, Atlas-like, the woes of the world on our shoulders. This book is about applied electronics as aids to security, where responsibility ends with providing to the user information in the most usable form – to the effect that all is well, or that an incident requiring attention is developing or has occurred. But it is very much our responsibility to sense what is going on. Is there a gun in that suitcase? Has she paid for that garment? Is there an intruder in the laboratory? How did he get there? Could he have been spotted before he got in?

Technical responsibility for sensing is limited only by the operational requirement and cost, as confirmed in the order for implementation, and here lie opportunities for us to excel and to compete successfully for business.

Problems arise operationally, legally and in insurance if a sensor fails to detect what it was designed to detect. Responsibility has therefore to extend to reliability and fitness for purpose. The surveyor has the responsibility of identifying and specifying action against evasion by the villain, to the extent at least that he cannot be charged with negligence. But security measures themselves inevitably have economic and operational limitations which defy precise definitions. Consequently, the limitation of responsibility rests heavily on the legal concept of what is reasonable in the circumstances. This theme is developed in Chapter 3. Likewise, the remainder of this chapter is preparation for more detailed treatment of 'thinking security' in later chapters.

Defining the problem, seeking solutions

If your business is the fabrication and stocking of copper products, it is fair to assume that you are highly skilled in all aspects of the business. It does not follow that you would be equally successful if suddenly you switched to running a business as a haulage contractor. Similarly, it may be unreasonable for you to know enough about security to protect your stock and business adequately. As in switching to haulage, so in considering security: normal common-sense will get you a long way, and ideas for solutions will start coming to mind quite readily. But solutions to what? The problem, of course: but do you know what the problem really is?

An example may help to illustrate the point. A colleague of mine maintains that finding solutions is easy and that the main need is to define the problem effectively. It may be of no use, for instance, for a Chief Security Officer to report to his director that x bottles of whisky were stolen yesterday, and leaving it at that. The likely reply by the director is that it has happened before and what are you going to do about it? Without solving the problem, quick remedies would have been to put in closed circuit television (CCTV) for a daytime loss and a burglar alarm to cover a night-time loss. As it happened, our Chief Security Officer was faced with daytime losses and, concentrating on this fact, he decided to trace through every step in the chain from executive decision to purchase more stock, through loading, transport and goods inwards for whisky received, and claims on insurance for shortages. It was found that systematic information on dates, times and quantities, of despatch and arrival, was being issued to personnel who had no 'need to know' the information. By discreet arrangement with the departments concerned, distribution of the information was restricted and made less specific, with the result that the losses ceased and the culprit was isolated. As my colleague would have said, the solution was easy, but the hard work went into discovering what the problem really was.

If the right answer to a problem does not involve the use of electronics, so be it: there is no obligation to use electronics when the job can be done better in another way. Back, then, to your copper business. It is no use jumping to quick conclusions and solutions when you suspect you have a security problem. Bring in instead a man whose daily job it is to think security, either your own security officer or an outside consultant, and give him authority to go where he has to and to get answers to his questions.

Preparing proposals, authority for implementation

Put yourself now in the place for a while of the security adviser. The investigation or survey has been carried out, the problem has been identified and provisional solutions have been drafted. Your next problem is to get your proposals implemented. That means persuading someone to spend money that probably has not been allowed for in the budget. In other words, you have to sell the proposition. And what is it that you have to sell? It is little more than an idea, the successful outcome of which is that nothing untoward will happen. People talk of the problems of selling intangibles – they should try this!

The hard sell may be effective in the domestic world, but in the business and professional world the likely effect is for the salesman to be shown the door. Skilful sales people are great believers in generating mutual trust and confidence between salesman and prospective customer, and nowhere, I think is this more true than in the security world, for if you take away mutual trust between people dealing with a security problem, what have you left?

With this background of understanding, it is perhaps easier to see that proposals need to be prepared and presented in factual and logical form with supporting narrative, rather than as a verbless tabulated list of equipment. The narrative should bring out the benefits of doing things in the way you propose, without being too lengthy, as the customer is all too anxious to get to the bottom line to find out what it is all going to cost.

Bear in mind also that your opposite number representing the customer may not be the one to authorise expenditure. Thus, your opposite number also has a selling job to do, within his own organisation, and what you really have to do is to provide him with the ammunition he needs to get approval for the project. This attitude helps to get you both on the same side, with perhaps the chief accountant as the common target.

One ploy that has worked more than once is to invite the chief accountant to join a walk round the risk. It is a new experience for him, he may be flattered by the invitation, and certainly he will have been unaware, till shown, how exposed his organisation is. In such ways the decision-makers can be persuaded to think security.

Insurance and police

When the first Police Crime Prevention Officer was appointed, my reaction was 'aren't they all – why a new function?' Sir Robert Peel founded the Metropolitan Police in 1829, with particular emphasis on crime prevention, but human nature is at work again, and policemen see promotion coming more easily from catching villains than from saying, 'Look, I am an efficient policeman – nothing has happened on my patch'.

By removal of the duty of catching criminals from the new function, the Crime Prevention Officer comes into his own as security adviser from the police point of view to the public at large. Security officers and surveyors need to know the Crime Prevention Officers in their area for advice and for independent support in getting security measures adopted. To maintain independence their advice has to be on principles and examples of applications of principles without being too specific regarding individual suppliers of equipment.

As we saw above, insurance companies have their own ways of providing pressure and support, and their attitude is totally single-minded in wanting to prevent a claim being made against them, conditioned by competing insurance companies and their need for business.

It seems inevitable therefore that each source of advice thinks security from its own point of view – and in writing proposals the security officer or surveyor has to draw upon his own experience to decide whether to compromise and retain support of various advisers, or to be single-minded and risk losing support from one or more quarters.

Security and safety

Moral and legal obligations regarding safety are increasingly influencing the way we think security. At one time it seemed that the majority of attacks occurred during the silent hours; and when premises are empty of people, there is hardly a safety problem to consider. Whether it is right or not to conclude that the anti-crime measures taken, physical and electronic, have been sufficient to deter night attack, it is a fact that day attacks have increased substantially, and the pattern of security measures is having to be developed accordingly.

As a case in point, attacks on banks are increasingly in daylight, presumably because the contents of strongrooms are becoming too difficult to get at successfully and without detection at night. The motives for attack, which now include getting funds to buy arms for terrorist activities, make villains willing to take greater risks, which they offset by carrying firearms. In the event of an armed attack on a bank during working hours, it would be easy, physically and electromechanically, for the bank to have its street doors fitted with automatic closers and locks, so that intruders could be trapped inside until caught by the police.

But the fire and safety people have the law behind them in a way that is denied to security people. If there is a risk that customers and staff may also be trapped inside the bank, then requirements may be imposed giving facilities for emergency release of the doors, for use, say, in the event of fire. The same facilities can be used by the villains to get out, and provision should have been made in the system design for an alert to have been raised when the doors were locked at the start of the emergency. If the emergency release facilities are used by the villains to get out, then the alert should have given the police sufficient time to prepare a suitable reception party for the villains as they came out.

In such ways can security be maintained, but the problem posed to security by the legalised emphasis on safety is acute. It demands the closest co-operation with the safety authorities to ensure that sufficient mutual understanding can be developed for the discretion allowed in law to the safety people in implementation of their measures to include adequate provision for security. In other words, they, too, need to be encouraged to think security, just as they expect us to think safety. And it can be done.

Discussion points

Do you always lock your home or your car when you leave it? If not, it may be wise not to confess the omission to anyone. It would be wiser still to resolve always to lock up in future. But what is the use of that if the locks can be 'picked', even by an amateur villain? Can first-class locks all round be justified? Have you thought of internal bolts for all but the final exit door? Have you started to notice lapses in security awareness in those around you? Have you grasped the significance of the 'dual question'?

Find a like-minded person you can trust to act as a sparring partner, and draw out of each other the reactions to questions such as those above, and to situations involving 'Did you notice that?', until you feel that thinking security is becoming second nature.

3 Security system concepts

So far we have been setting the scene, by getting our minds in tune with security, and finding out how the book proposes to deal with its subject of 'Electronic Security Systems'. Now it is time to get down to sorting out what security systems are and the part played in them by electronic aids.

Objectives and motives

It does not seem so many years ago that the only time a security system was called for was when a customer could not get adequate insurance cover without one. That is still very much the situation for burglar alarms installed indoors. The growing need for outdoor detection has other origins.

One does not have to look too far for evidence that there are forces at work that are dedicated to changing the existing order of things. At one time it would have been a fair guess to suppose that the proceeds of a bank raid would be spent in high living in the South of France or wherever, but today the objective is at least as likely to be to buy arms, information or people. We have entered an era in which the intruder is trained, is intelligent, and may be motivated to damage or destroy the target country's economy and way of life through subversion, intimidation and sabotage.

A sombre thought? Maybe: but if things are quiet, it is because our political, military and civil defence forces are being successful in maintaining that position. And that explains the need for our services. Our purpose is, with others, to preserve the *status quo*. If nothing happens, we have won. The policy most likely to achieve this objective is that of deterrence – that is, creating in the mind of the villain the belief that he cannot succeed.

If we say that our policy is to deter, our weapon is time. If in spite of our efforts an attack does develop, then physical protection and delay are given by walls, fences, window bars, and the like, and warning that an attack is developing is provided by the intruder detection security system.

It is necessary to note also the two operational elements in security – the active part played by manpower and the passive part played by the physical and technical aids to that manpower. Any time gained by the detection system can be used to bring the protective reaction forces into action earlier.

Assessment of intrinsic and consequential risks

So far we have tried to envisage the problem by outlining motives for attack, and looking very briefly at the political, economic, technical and psychological aspects of defence. As a good friend often reminds me, the more accurately you can define the problem, the greater are the prospects of finding a good solution.

I do not know how anyone can assess a physical risk without going to see the area in question and its environment, and I made it a rule not to make recommendations on a security system without first doing just that. Only after doing this can you ask the right questions of the people concerned.

To take a simple example: your escort will readily take you to a central heating contractor's warehouse where valuable stock is kept, and expect you to design a security system to cover them. Well and good, but he is less likely without being prompted to brief you on where and how the keys are kept, or where the telephone cable runs. The consequences to the smooth running of a business of loss of keys and of having the telephone cable cut can extend far beyond the immediate problem of, say, material loss and having the alarm system put out of action by an intruding villain.

But one has to accept that there are times when the customer considers the risk is such that no one but the successful contractor should be allowed to visit the location or to meet him directly. All negotiations have therefore to be conducted through a third party. On one such occasion my client lost the contract to a competitor who appeared to satisfy the third party that he could meet the specification. When the competitor eventually met the customer, he soon realised that the real risk was not fully covered by the specification and that he was unable to comply. I only know this because the competitor got in touch with me to find out whether my client could comply with the real, as distinct from the specified, requirement. Fortunately, he could.

I do urge those who believe they have high risks to cover, to meet a short list of potential contractors and to check their credentials. If there are grounds for mutual trust, then the potential contractors should be allowed to meet directly with the customer so that the implications of the requirements can be understood. Only in this way can the embarrassments of the above example be avoided, and, more important, only thus can the customer be sure that he has done his best to obtain the security system he needs.

If the risk is such that the customer still has to say 'No', then the specification has to be prepared by someone who understands both actual and consequential risks. Failing that, he may find that potential contractors also say 'No', and the customer may be better advised to do it himself. Some do.

Matching systems to risk assessment

The law of diminishing returns applies to this work as to any other. The simplest of systems is indicated for the small retail shop requiring an installation to satisfy an insurance requirement. As the value of the risk increases, so does the range of choice in system design. Emphasis on gaining time tends to concentrate attention on the wall, doors and windows, but a sufficiently impregnable perimeter will lead a determined intruder to consider methods of evasion, such as finding a way through the roof.

We are concerned, therefore, not only with detecting the intruder at the outer perimeter, but also at various stages in his approach to his target. Rather than spend our all on, say, the perimeter only, resources must be properly balanced to give second and possibly third lines of defence.

The value and limitations of physical protection

The physical, or fortress, concept of protection has been with us from the earliest times, and consequently its value is the most readily understood of its properties by the public at large. The ultimate in their eyes is the present-day Fort Knox, which epitomises all that a stronghold should be. But who but the USA could afford such a structure? If to provide a company with physical protection would cost them their profits for the next ten years, would they buy? The answer probably is 'No', and herein lies one of the limitations of physical protection. If companies cannot afford 'perfection' at the price of ten years' profits, they will ask, 'What can you do for less?' To provide anything less than perfection means taking a security risk, and you will find that that runs through the whole of our business – security is a risk business.

The security risk with physical protection of anything less than perfect quality is that it can be penetrated – sooner or later. And the further problem is that no one might know of the penetration until it was too late, until the valuables had gone.

As we shall see in the next section, the obvious answer these days is to equip the stronghold with electronic detection aids to give warning that an attack is being made. But these, too, have their limitations, the pendulum swings, and we are again saying, 'Don't relax on physical protection, just find the optimum combination of physical and electronic cover to minimise the risk of loss'.

The functions and limitations of electronic detection

The operative word in this section is 'Detection'. All right: electronic systems can and do provide protection in the sense of deterrence, of generating in the mind of the villain the fear that he will be caught in the act. But to be caught in the act, that act has to be detected, and information has to be transmitted to people, and the people have to get to site in time to catch the villain. In spite of the problems, this is complete justification for the use of electronic detection. A quick warning of attack must be better than not knowing even that an attack was taking place. Electronic systems are rightly called 'aids' to security personnel, and in discussing proposals it is generally wise to ensure that the distinction between protection, as with reinforced walls, and detection, as with electronic sensors, is understood and accepted.

Electronic detection is relatively inexpensive, and the only limitation of a properly designed and installed system is its dependence upon people. The system can give its warning, but it is dependent upon people being there to accept the warning and upon people being available to act upon the warning, in time.

A useful exercise in the Discussion Points of this chapter would be to consider how limiting, if at all, this problem is and whether anything should be done to combat the trend again towards greater physical protection, arising from customers' belief that they cannot rely upon sufficient police or other reaction forces being available to deal with their own particular attack when needed.

Alarms for perimeters and interiors of buildings

If electronic security systems are to be effective aids to security personnel, then the communication link between the aid and the personnel is crucial. Colloquially this is recognised by calling it an 'alarm' or, more specifically, a 'burglar alarm'. In the larger risks, and in the professional security market, the intruder's intention may not be burglary but espionage or sabotage, and the tendency is to call systems for detecting his presence 'intruder alarms'.

In the public mind the concept of a burglar alarm is embodied in the tinkling bell outside a shop in the High Street. Sadly, the fact that they go on tinkling discredits the concept. 'No one takes any notice of them' overlooks the fact that the very person who matters in this case, the villain who has accidentally triggered the alarm, has noticed and in all probability has gone. Alarm companies recognise this by putting in a timer unit to silence the bell after a reasonable period.

In addition to scaring the villain, the function of the external alarm bell is to attract the attention of patrolling police or, hopefully a member of the public, who would report it to the police for attention.

Not surprisingly, the prospect of the police not being there to notice the ringing alarm bell, and of the public ignoring it, led to concepts of more direct signalling to the police, using automatic calling on the telephone line network. Naturally, the police said that if they were to accept burglar alarm calls they wanted a chance of catching the villain, and accepted automatic signalling provided that there was no local alarm bell to scare their man away. Conversely, the insurance companies objected, saying they wanted to get rid of the villain as quickly as possible, and the bell must stay.

Inevitably, compromise was reached, with the alarm bell being kept silent for long enough for the police to get to the premises, and then to ring in case the police didn't arrive. These and the more advanced aspects of alarm signalling are examined in Chapter 8.

Whatever the method of alarm signalling used, the police of various countries say that about 90 per cent of the calls they receive are false. By false they mean that the call was not originated by a villain. This is an appalling record for the security industry, and self-induced and external pressures to improve the situation have to be welcomed. But pressure is no use without awareness and understanding of the nature of people and of the electronic security systems that give the trouble, and that is a major theme we are trying to cover in this book. The overriding consideration in designing, selecting and installing detection sensors, control equipment interconnection and signalling systems for buildings has to be reliability and avoidance of false alarms. Many aspects of false alarms are covered in various parts of the book, and the index will help in putting together a comprehensive reading list of references to the topic.

Systems for outdoor perimeters and areas

Let us look for a moment at the changing nature of typical operational problems. Imagine, if you will, a night watchman huddled round his brazier. He was glad of the job no doubt, but his employer was no philanthropist. The very presence of our watchman did have a deterrent effect at least upon vandals, and if anything more serious did start to develop, he was expected to call 'stop thief', and this was about the extent of his duty, and of the risk.

Security system concepts for today have to be more complex. The risk areas may be factory sites, military establishments, Government buildings, public services, storage depots, and so on, mainly sites likely to justify already the employment of the modern equivalent of the night watchman, round-the-clock security staff.

Besides intrusion, security people have to cater also for 'extrusion' – that is, the departure of a stay-behind employee, collusion between an employee inside and a colleague outside, and the ever-growing skill used in prison escapes. It is no longer practicable for the site security personnel to monitor their area unaided, and this need has led to the intense development of electronic aids.

In the previous section mention was made of the false alarm problem. Out of doors it is still more difficult to avoid false alarms, particularly on perimeters where a wall or fence, for instance, marks the barrier between the site and friend or foe alike outside. Discrimination between the two becomes the challenge for the system designer.

Weather and other environmental effects add to the difficulties of discrimination, and it is rare that the false alarm rate can be made sufficiently low for an outdoor detection system to be added to an unmanned site, or to replace manpower on a site completely. Nevertheless, the rising cost and falling availability of suitable manpower increasingly focuses attention upon improved physical protection backed up by electronic detection at the outer perimeter, at intermediate points within the sites and on routes that must be taken by our intruder towards his objectives. Chapters 6 and 7 are devoted to aspects of this subject.

Systems for use within working hours

The concept of a security system in the mind of the general public is a system that is used in lock-up-and-leave premises – in other words, for use outside working hours. As we have seen, it may be that the very success of security systems for the silent hours has shifted attention by the villain, and consequently by security people, to working hours activities and systems.

An obvious step is to keep the doors closed and locked. But working people have legitimately to move to and fro with relative ease in order to do the work they are employed to do. Thus, again, we have to discriminate between friend and foe, preferably without having to send someone to the door each time to check, to accept or reject, to open and close and lock the door for each personnel movement. Out of this problem the concepts of door control, access control and surveillance have arisen, as enlarged upon in Chapters 7 and 10, in which the problems of people and what they carry will also be dealt with. Can we say for certain that people are not carrying weapons or stolen goods, hidden in clothing or luggage?

Components of typical systems

When studying a security risk, a surveyor normally has in mind the range of detection, control and signalling components stocked and used by his firm, and he selects from that range the components he thinks most suitable for the risk when preparing his system proposals and specification.

In this he is dependent upon his employers having selected a sufficiently wide range of components to cover every reasonable contingency. A favourite question of mine is, 'What do you want that doesn't exist?' Socially it is a conversation stopper, but in training for security it really points the way. A surveyor is in as good a position as any to understand and answer this question. If he has assessed the risks competently, and thinks ahead, he will sooner or later meet a situation in which he says in effect, 'I wish there was a widget that would . . . '.

Out of little more than door switches and pressure mats in the early days have come answers to that wish, such as inertial switches to detect attempted evasion of door contacts, ultrasonic and microwave sensors of presence, 'sniffers' for explosives and drugs, infrared CCTV for seeing in the dark, microprocessors for voice verification, and so on.

An objective of this book is to encourage creative thinking. That it can be done is evident from the examples given, but it would not be so if security executives, surveyors and engineers passively accepted the range of components already available to them as the only ones with which to attempt to do their jobs.

Components and equipment will be examined progressively throughout the book, in reading which the dual question, 'Will it work? Can it be made not to work?' needs to be asked again and again to maintain awareness and to identify new ways of a weakness in any device being exploited by a villain.

Collective restraints on system concepts

However alert one is to the need for improvement, one is faced always with collective resistance to change and with the need to conform.

On the practical side of running a successful security business, it would not be feasible to allow each surveyor and engineer to do his own thing. In order to maintain proper control of costs and adequate profit margins, it is necessary for security systems to be designed round a predetermined range of equipment, components and working methods. With high labour costs, working methods may be the key factor, for it stands to reason that the cost of doing a given job the same way again and again on various sites must be lower than the cost of doing 'specials' on each site. It is the so-called 'learning curve' coming into play again.

Even when an employer has dealt successfully with economic constraints, he is still not a free agent to run his business in his own way. He will be restricted to the use of approved equipment for connection to telephone lines for alarm communication.

For him to be awarded the contract, he will have to be an 'approved installer' as defined by a National Inspectorate. In its turn the Inspectorate has to have a standard reference against which to inspect, so a further constraint is imposed by any National Standards that may be deemed applicable.

If he is to influence future standards, he is likely to join a Trade Association, which have their own conditions of membership. Where the police are the reaction force involved in case of trouble, they are entitled to, and do, impose restraints upon what form of alarm signalling and information they will accept and act upon. The ultimate sanction by the police is, in the event of excessive false alarms, refusal to act upon any further alarm calls from given premises, or even from a given alarm company's equipment.

System integration

Such, then are some of the range of constraints upon security system concepts. The surveyor, the system designer and the security manager need to make themselves familiar with these constraints, and more, by asking questions and trying to understand why the constraints are there. After analysis of problems comes integration, the putting together of the various ideas and devices needed to meet the security requirement.

In the chapters that follow we shall delve more deeply into system concepts and into the electronic equipment from which the armoury of a security system is built up. As reference to space detection occurs so often, perhaps we ought to devote a chapter now in an attempt to find a way of understanding space detection devices and methods. Because people cannot 'see' how space detection works, they tend to think that it is beyond their understanding. This is to do themselves an injustice, because it can be understood; read Chapter 4 and see for yourself.

Discussion points

On page 19 we encountered the pendulum problem of physical versus electronic security, related to the availability of reaction forces to attend

to any alarm signal giving warning of attack. Which way would you go –
to physical security, to more detection equipment or to more police able
to respond to an alarm call? Or would you tackle the problem at its roots
– the motivation of the villain?

Another exercise to work on with a like-minded, even abrasive,
colleague is to imagine what you would like to have that at present does
not exist. The exercise is just as valuable if you prefer to go outside
security.

4 Space detection fundamentals

At first sight, this chapter may seem to occur rather early in our book, so why is it here? It is included at all because in drafting the later chapters I found I had to use words that may be unfamiliar, or may have unfamiliar meanings in the security context. It is *here* because I have to start using some of the words in the next chapter.

Space – understanding something you cannot see

Space is what is left when everything else is taken out. Space is where a human being can go if there is nothing in the way. We are talking about ordinary space, where you and I and the villain can go – not outer space, where the astronaut can go.

If we want to know whether a human being is there, we can use electronic methods of detection of presence and of movement in space. In the security world space detectors go under various names, not all of them everyday words – microwave, radio, ultrasonic or infra-red: names which indicate rather the principle of operation of the detection device than what the detectors really do.

We have no readily understood word to describe what they do, what their function is, so a fair amount of mystery and misunderstanding has grown up around the whole subject. In this chapter I am going to try to simplify it all, to bring some understanding and perhaps to do a little debunking.

Now, you know what happens when people do not understand something – they tend to cover up and somehow they still display confidence, just like a child jumping into water or dashing across the road, not realising the consequences of what they say or do. In another context this is surely an illustration of absence of awareness. But in space detection there is no need to cover up: we are all in the same boat. How can anyone really understand something that cannot be seen? So relax while we try to relate space detection to a few things we are familiar with in everyday life, and even if we cannot understand just why it works, we can do our job if we are able to understand the words, what happens and what to expect to happen in differing circumstances.

Words

The sun is a *transmitter* of light and of heat. A piano is a *transmitter* of sound but not of light. A gas cooker is a *transmitter* of heat and of a very little light. A fluorescent tube is a *transmitter* of light and of very little heat or sound. A radio station is a *transmitter* of, not light, nor heat, nor sound, but we know that it transmits something – what? It is hard to know just what, but some clues are given further on. We say that these are transmitters, but how do we know? Only because Nature has provided, or man has made, *receivers*.

The eye is a *receiver* of light. The ear is a *receiver* of sound. The saucepan is a *receiver* of heat. A transistor radio is a *receiver* of what?

So far we have two words – 'transmitter' and 'receiver' – that have meanings in everyday life and also have meanings in the field of electronic security equipment.

Sound generation and reception

Play the note middle C on the piano. To the human ear it is a lovely rich sound. But also it is a string or wire, of definite length and definite tightness. It is also a hammer, made to move sharply into contact with the string when a finger presses the piano key. It is *vibration* which can be seen by looking at the string or felt by touching it. The hammer pushes the string to one side and stretches it a little, but as soon as it can, the string springs back again; however, like a pendulum, it overdoes it, and the string carries on to the other side and stretches itself again in the process. Back it comes again, and although the vibration becomes weaker (that is, the string doesn't travel quite so far to each side each time it moves), the time taken for the string to move from one side to the other and back to its starting point is the same. That is another way of saying that in any given time it will vibrate from one side to another and back again a definite number of times, and that number, if we say the given time is one second, is the *frequency* of vibration – another word we need.

If we play the note just to the right of middle C on the piano, the frequency of this new note sounds a little higher than that of middle C. Inside the piano we can see that the string is slightly shorter, and when hit by the hammer, it vibrates slightly faster. And so on, to the top note of the piano. Here we see that the string is shorter still; it vibrates much faster when hit, thus making the frequency of the note sound much higher. To get a scale of values, middle C has a frequency of, say, 264 complete vibrations per second, and the top note has a frequency of 4224 vibrations per second.

The word *cycle* was a very good word for saying one complete vibration, but it is traditional to honour pioneers of physics by giving their names to physical properties. So it is that the meaning of the words 'cycle of vibration per second' is given by the unit hertz (Hz), from the name of the pioneer, *Hertz*.

The same experience of vibration can be had via your hi-fi loudspeaker by playing, say, some organ music from your CD, and watching the loudspeaker cone. The lower the note from the organ, the slower the cone moves in and out. The devices required for converting the energy in organ music into a CD recording, and back again into sound are called *transducers*, which the dictionary says are devices that convert one form of energy into another.

Ultrasonics

In the piano and loudspeaker analogies, we have been dealing with sonic or audible sound. But it is reasonably well known that dogs and other animals can hear higher sound frequencies than can be heard by humans. Given a suitable transducer, one can readily imagine a frequency or note twice as high as the top note on the piano. Humans can hear that too. Now double it again to 16 896 Hz and we are just about on the upper limit of human hearing. Go a little higher to, say, 20 000 Hz. A dog can show that he can still hear it, reassuring us that it is real, although we cannot hear it ourselves. From this frequency upwards we refer to *ultrasonic* sound energy.

Now let us go back to the piano for a moment to find another important word. We noticed that the higher the note or frequency the shorter the string, and the lower the frequency the longer the string. The string is there to produce a musical note for us to hear, and it does that by making the air surrounding the string, and between the string and us, vibrate at the same frequency. Is it stretching the imagination too far to suggest that as the string vibrates, it first pushes and then pulls the air immediately surrounding the string? There is a knock-on effect on the air just beyond the string, and the successive increase and fall of air pressure spreads out rather like the waves from a stone dropped into a pond. The distance between the crest of one wave and the next in water, or between adjacent high-pressure points in air, is the length of the wave, or *wavelength*. But there is no water flow, or air flow: only the wave travels.

Relating frequency and wavelength

We have enough words now to look at three rather remarkable facts. We saw earlier that the higher the frequency or pitch of a note the shorter the string, and the shorter the wavelength.

Fact 1 is that if we multiply the frequency of any note by the wavelength of the same note, the answer is always the same – a constant.

Fact 2 is that the constant is not just an otherwise useless number: it is the speed at which sound travels in air, that is, 332 metres per second.

Fact 3 is that if we know the frequency of any note in air, we can find its wavelength, because by combining Facts 1 and 2 we have

$$\text{frequency} \times \text{wavelength} = \text{constant}$$

or

$$\text{wavelength} = \frac{\text{constant}}{\text{frequency}}$$

As an example, we find that the wavelength corresponding to 1000 Hz is

$$\frac{332 \text{ metres per second}}{1000 \text{ Hz}}$$

$$= 0.332 \text{ metres}$$

Or for those who like to think in feet, the wavelength of a 1000 cycle or hertz note is just over 1 foot.

It is handy to remember these figures as round-figure ready reckoners. For example, the wavelength of a 100 Hz note must be just over 10 feet or 3⅓ metres long, or an ultrasonic frequency of 30 000 Hz has a wavelength of about 11 mm.

More words

The background to space detection is now beginning to take shape and it is worth being patient until further words that have to be used have been shown to have meanings that can be understood and accepted.

The next two words are simple enough. We say we have an *echo* when we hear a sound *reflected* back towards its source. We hear first the sound of a rifle shot direct from its source, and then we hear the sound again after reflection from the solid safety barrier behind the targets.

Doppler

Imagine now an empty room about 20 feet square and 10 feet high, in which there is no movement at all, not even air movement. In one corner of the room about 6 feet from the floor imagine an ultrasonic transmitter, similar perhaps to a loudspeaker tweeter, and beside the transmitter an ultrasonic receiver, both facing towards the opposite corner of the room. The transmitter and receiver arranged together in this way form an ultrasonic *radar detector*.

The transmitter will be sending out ultrasonic energy at something above 20 000 Hz (or 16.6 mm wavelength) and the receiver will receive energy at the same frequency, some of it direct from the adjacent transmitter and the remainder reflected from the walls, and maybe from the floor and ceiling. Next, imagine that there is a door in the corner of the room facing the ultrasonic transmitter and receiver, and that an intruder comes through the door into the room. Some energy from the transmitter that would have been reflected from the wall and closed door is now reflected from the moving intruder back to the receiver. The crucial question is whether the receiver can tell the difference between signals from fixed and moving objects.

The answer is that it can. We saw earlier that

$$\text{frequency} \times \text{wavelength} = \text{speed of sound}$$

Now, as the intruder moves forward, the energy reflected from him gets back to the receiver early, 'before it was expected'. In other words, the receiver thinks momentarily that the speed of sound has increased, making the equation

$$\text{frequency} \times \text{wavelength} = \text{speed of sound} + \text{increase}$$

That cannot be right, however, as the equation is now unbalanced. As we said that the air in the room was still, nothing has happened to change the wavelength. So the only thing that can happen to make the equation true again is for the frequency to appear to increase. We then have

$$(\text{frequency} + \text{increase}) \times \text{wavelength} = \text{speed of sound} + \text{increase}$$

And this is just what the receiver notices – an increase in frequency of reflected signal from the intruder. The increased frequency is compared electronically with the fixed transmitted frequency, and the difference is used to trigger an alarm signal.

Something similar happens when we use our ears as receivers, and notice the change in frequency as a car sounding its horn passes us in the street. The scientist who first explained this acceptably was *Doppler*, and we now speak of the Doppler shift of frequency, or the Doppler effect, or just Doppler.

Attenuation

A fact that runs through Nature is that the higher the frequency the greater the *attenuation*, or loss of energy. Put your hands over your ears while listening to music. All the sounds will be reduced in volume, but notice that the top notes, if audible at all, are now much quieter than the low notes. The same happens if a loudspeaker is placed behind a heavy curtain.

This applies to sound waves in air, but Nature is just as consistent for light. Remember the spectrum of colours transmitted by the Sun which together make daylight: the violet end of the spectrum has the higher frequency and shorter wavelength energy, and the red end of the spectrum has the lower frequency and longer wavelength energy. At mid-day we expect normal daylight, but at sunset the light has to travel obliquely through much more of the Earth's dusty atmosphere. The atmosphere attenuates or cuts down the violet and blue higher frequency light, and we are left with a red-looking sunset – or sunrise, for that matter.

Nature's consistency

Nature extends its consistency to include heat, radio and microwave energy with light, all being transmitted through space by *electromagnetic* waves. In fact, heat, light and radio are all electromagnetic waves, only the wavelength is different. All are attenuated to some degree, as we saw in the sunset example. Instead, however, of being restricted in speed through having to push air to and fro, as happens in generating sound or acoustic waves, electromagnetic energy meets virtually no restrictions to its speed. In fact, it is the fastest thing there is, and travels at 300 000 000 metres per second, or, in round figures, at 186 000 miles per second, quite a contrast with sound travelling in air at 332 metres per second.

Even more astonishing is that Nature retains its consistency in such a drastic change of transmission medium as that between air and space by retaining the relationship we found for sound:

$$\text{frequency} \times \text{wavelength} = \text{velocity}$$

More on attenuation

To illustrate how attenuation increases with frequency, we have used analogies from both acoustic and electromagnetic sources of energy. Fortunately for us in intruder detection, we are concerned with tens or, at most, hundreds of metres of detection range, which are short distances compared with the distance light and radio waves normally travel. Although these waves can be shown to be attenuated increasingly as their frequency increases over long distances, loss of energy due to distance travelled is so very small for radio waves over the distances we are interested in that radio attenuation can be neglected in design and use. It doesn't matter.

For ultrasonics, however, increasing attenuation with frequency does matter, and this limits quite severely how high in frequency we can go before so much energy is lost that not enough is left to give a usable echo from the intruder. As a guide, power loss is roughly proportional to the square of the frequency. So, doubling the frequency from 20 000 Hz to 40 000 Hz would reduce the echo power to a quarter for the same transmitted power.

Dispersion

Now we shall look at other fundamentals that limit the working detection distance of space detection devices.

The inverse square law

Imagine a slide projector for viewing photographic transparencies. If the screen is set up at a distance of 1 metre from the projector, with no slide in the gate, a square patch of light will show on the screen and the brightness of the light on the screen could be measured. If the distance between the projector and the screen is now doubled, to 2 metres, the patch of light on the screen will increase in size. Measurement will show that the vertical and horizontal sides of the patch have doubled in length, so that the area is now four times the original. The power of the projector lamp, however, is still the same, so the light from it covers four times the original area.

Common-sense reasoning says, therefore, that by doubling the distance between the projector and the screen the brightness of the light on the screen is reduced to a quarter of the original. The same reasoning says that if we want to maintain the same screen brightness after doubling the distance to get a larger picture, we would have to increase the power of the projector lamp by a factor of four from, say, 100 watts to 400 watts.

This relationship is called the *inverse square law*. The relationship applies equally well to radio, microwave, passive infrared and ultrasonic intruder detectors, but when the transmitter and receiver are side by side, as for the radar type of sensor, the relationship takes on an ever-greater significance, as explained below, and further in Chapter 15.

The inverse fourth power law

Given that we are considering a radar-type intruder detector, the energy is transmitted and dispersed in proportion to the inverse square law. The same law applies to the energy reflected back from an intruder to the receiver beside the transmitter.

To use the slide projector analogy again, and assuming near-perfect reflection of light from the screen, the screen itself becomes the transmitter and the eye beside the projector becomes the receiver. Now, if we succeeded in keeping the screen brightness the same after doubling its distance from the projector, the eye will again see only a quarter of the original brightness, because of the inverse square law. In all, the inverse square law applies twice – once for the light from the projector to the screen and once from the screen to the eye. The brightness at the eye is one-quarter of one-quarter, or one-sixteenth, of that seen before we doubled the projector to screen distance.

Happily for human beings, the eye adjusts automatically to compensate for much of this change, but a radar sensor has no such adjustment. Radar receivers are usually working near their maximum capability most of the time, while transmitter power is limited by economics or by Government restrictions.

Just to get the scale of values firmly in one's mind, in order to double the detection range of a typical radar intruder detector, it would be necessary to increase the transmitter power by a factor of 16. As this cannot normally be done, some people increase the receiver sensitivity. By doing this they increase the receiver sensitivity to any interfering signal and so risk aggravating the false alarm problem.

Beam shaping

There is one more fundamental to be understood if the use of space detection is to mean what it should. Using the slide projector again, suppose that the projector filled the screen when 10 metres from it. And suppose that there was no room for the projector in that position and that it had to be put at the back of the lecture theatre, at 20 metres from the screen. We have seen that the picture would now be four times the size of the screen and that the screen brightness would be reduced. The projection-ist can retrieve the situation by changing the projector lens to one having a focal length of twice that of the original lens. If the beam angle of the original lens was, say, 40° vertically and horizontally to fill the screen, then changing the beam angle to 20° by doubling the focal length will again only just fill the screen from the back of the theatre and the screen bright-ness will be maintained.

Beam shaping in this way applies equally to space detection devices. Taking the radar example earlier, we found that we would have to increase the transmitter power by a factor of 16 to double the maximum detection range on an intruder. If, however, the beam angle of radiated and received energy is reduced both vertically and horizontally from, say, a normal 80° to 40°, then the effective detection range can be doubled without increasing the transmitter power or receiver sensitivity. This is *beam shaping*, a technique used widely in radar using reflectors, horns or lenses, and could be used more effectively in intruder detectors if better understood.

Being at home with space detection

Once you realise that there is not much difference between things you know and understand and the way various space movement detection devices work, you should have no difficulty with later chapters.

If you feel at home with photography, or with heat, light and sound in school physics, you can feel at home with microwave radar sensors. It is the same thing, using electromagnetic waves, only the wavelength is differ-ent. If you feel at home with microwave radar, you can feel equally confi-dent about ultrasonic radar: the wavelengths are much the same; only the transmission medium is different – air instead of the ether. And passive infrared fits comfortably between light and microwaves.

More on practical properties and use will come in later chapters, but meanwhile try using the following discussion points to check how much you have absorbed from this chapter.

Discussion points

(1) Why are electronic flash still photographs often disappointing?
(2) Can you hear sound having a wavelength of 10 mm? What is the wavelength of the highest frequency sound you can hear?
(3) Which has the greater attenuation – microwaves or infrared? And why?

5 Surveying for intruder detection indoors

By indoors we mean in buildings, including their outer walls, inner walls, doors, windows and other apertures, floors, ceilings and roofs, and the space within them. An area of ground surrounded by buildings, a courtyard, is not regarded as indoors.

So now we are carrying on from where we left off in Chapter 3. There we talked of security system concepts, and here we shall consider how a requirement for intruder detection indoors comes into being, and what follows from that requirement.

Origins of requirements

This is important. For most of us a requirement appears as a verbal or written instruction to do something. That instruction could come from our boss, but, with respect, let us look behind him for a few minutes. It is not enough now to say it originates with an awareness that a risk exists, although awareness is the fundamental ingredient. It comes from awareness that physical protection is not enough, and awareness that a warning is required that the physical protection is being penetrated. So who is it that is aware of these things?

The *insurance underwriter* has the closest interest, as it is he who accepts liability to pay up in the event of a claim for any of a specified range of losses occurring. Before accepting potential liability, he may call for an opinion regarding the risk from the insurance company's local inspector, who may not be specifically trained in this form of assessment, or from the *insurance surveyor*, who is more likely to be trained and qualified not only in assessing the risks, but also in preparing a specification of what physical protection and what form of intruder detection would be necessary to make the risk acceptable to the underwriter.

The *Police Crime Prevention Officer* is another source. Where sections of the civil police are detached from the duty of arresting villains, they are more free to visit new premises, existing ones undergoing alterations or change of use, ownership or tenancy, and premises that have recently experienced a loss. They make recommendations on the security measures that should be taken to protect the property, and advise on how the precautions should be carried out to facilitate both routine police surveillance and also police emergency attendance in the event of an attack.

The *industrial security officer*, evolving from the early night watchman concept, has become a highly trained and experienced civilian officer on the payroll of industrial, Government and Government-controlled organisations having risks that cannot reasonably be expected to be covered adequately by reliance upon the civil police forces. With their own uniformed staff they come close in effect, if not in fact, to being private police forces. The risks involved typically require continuous round-the-clock monitoring, as distinct from lock-up-and-leave type risks. The Chief Security Officer and his staff are frequently the source of electronic security system requirements.

Departmental Managers. One may not be conscious that local managers are sources of security requirements, but in organisations where total responsibility is delegated to the departmental managers for the efficient running of their departments, an acute awareness of the possible causes of loss develops, including losses that come within the sphere of security. Rarely does the local manager deal with these himself – he has to bring in the Chief Security Officer or equivalent to help him, who, in turn, becomes the visible source of a requirement.

Security Company Initiative. Like the radio and television markets before them, there was in the early days more than enough business to go round the firms selling a security service. Business came to them by letter and by telephone from contacts established within the circle of sources touched upon above. These contacts had and have to be cultivated and established as much on grounds of mutual personal trust as on grounds of specific products and services.

The 'it will never happen to me' attitude will ensure that the market never becomes saturated, but for those who did and do take security seriously the need becomes increasingly one of replacement and extension to existing systems. The people one knows who were effective in establishing new systems seem less effective in initiating replacements. If the original system has been effective as a deterrent and no attacks have occurred, why bother to replace it?

Thus, it falls to the security companies themselves to take the initiative – to sell the idea that the villain is learning, and that he now has new motives for attack, and that the old systems need to be replaced by their modern equivalents.

To get round to all former customers on this basis compared with dealing with their original contacts is a formidable business for security sales staff. And generating new business, as in the domestic house market, is even more daunting. No wonder, therefore, that the one-time unrewarding national and local advertising method is finding favour on an ever-increasing scale. And the theme? 'Fear'. Fear of the consequences of not taking security precautions is now very much more easily worked upon, since, as we saw in the opening chapter, statements such as 'Security was very tight' are so widely used by the media when reporting national and other events.

Our aim is crime prevention. Are we justified in using emotions such as fear to persuade people to do the things necessary to bring peace and peace of mind? Try this as a discussion point later, not forgetting that if they are to stay in business, making a profit on the transaction also has to be one of the aims of the security industry. If in discussion you find the second aim distasteful, consider also the consequences of the security industry going out of business.

Site surveys

I said earlier that I would never make recommendations regarding the security of premises without first seeing them. Equally important is talking to the people concerned. Beyond the deterrent effect, no amount of electronic security is any use against the determined intruder unless the system includes appropriate equipment for informing the right people that intrusion is taking place. This is a statement of fact. Practically everything else about a site survey is a matter of opinion.

That is sufficient reason to encourage each individual concerned with carrying out site surveys, or with managing surveyors, or with training surveyors, to work out a guide procedure to be followed in surveys. In this way the element of opinion is acknowledged and yet it allows codes of practice and policy guidelines to be incorporated.

If we cannot keep an intruder out of the premises physically, the purpose of the survey is to establish three things:

(1) In what way may an intruder attempt to achieve his objective? His objective is the risk, and getting clear on what this is and the way the intruder may reach it is *risk analysis*.
(2) From the risk analysis the surveyor or a systems engineer can work out what measures, including the provision of electronic equipment, are needed to warn those concerned that intrusion is taking place. Working out this is *system design*.
(3) The engineering department has to work out how to implement the system design, deciding in detail equipment selection, cable routes antitampering provisions and the finer points in alarm signalling. This is *installation design*.

In these three paragraphs we have said what, what, how – a useful way of remembering the purpose and outcome of site surveys. But whichever aspect you are dealing with, the co-ordinating feature to remember is that intruder detection is a communication problem. A survey is used in what follows as an introduction to the various elements of an indoor intruder detection system.

Perimeter detection methods

Those who are relatively new to intruder detection may find it logical and easier to follow if we imagine a simple survey, and start where a friendly visitor or an unfriendly intruder might start – that is, at the perimeter of the building. So I will ignore my own advice for the time being and ignore the alarm signalling system.

Question 1. Is the intruder detection system needed for use during working hours, or during the night and weekends when the premises are unoccupied?

Straight away we find the need for an escort – someone who represents the customer, the owner or manager of the premises, who is familiar with how the building is used, and who is authorised to answer or to get answers to all reasonable questions the surveyor may put.

To get a stranger to reveal his security problems to a stranger is not the easiest of jobs, and the ability of the surveyor to establish an atmosphere of mutual trust with the escort goes a long way towards designing an effective system.

If the intruder detection system is needed during working hours, then Chapters 10, 23 and 25 on door and access control would be more appropriate, but if the system is needed for the silent hours, read on.

Question 2. When the premises are being vacated at the end of the working day, which door is used as the final door of exit?

Go and look at it. It might be a front, side or back door. The location doesn't matter too much, provided that it can meet needs such as:

- allowing the keyholder the safest route to and away from the building
- resistance to being kicked in – preferably outward-opening with tamper-resistant hinges and a frame securely fixed to the walls
- having a security type lock activating one or more deadlocking bolts.

Question 3. What other perimeter doors are there?

Keeping in mind the final door of exit, each of the other perimeter doors should be examined in a similar way, particular attention being paid at this stage to the physical defensive properties. Substantial internal bolts at the top and bottom of doors should be a minimum requirement. If one of the other perimeter doors seems to be more suitable as a final door of exit, discuss it with your escort. If he doesn't agree readily and you still feel strongly on the matter, you can perhaps discover his true feelings by asking whether he would object to your writing your suggested alternative exit door into your proposals. Whatever technique is used in finding out what he is thinking, remember that it is likely to be your escort who will be selling your proposals, or those of somebody else, to his boss, and that it is worth making every effort to arrive at consensus during the personal and verbal stages of a survey.

After allowing for any necessary physical improvements to the perimeter doors, we still have to face the prospect of a villain trying to get through one of them, and we need to know when he does so. What detection devices are there that would trigger an alarm signal?

Concealed wire and foil

The photograph in Chapter 15 of the hole made in an art gallery door illustrates the problem. A long-established and simple precaution is to cover the inside of the door with a length of thin-gauge wire or foil strip in a regular or irregular pattern, so that cutting a hole through the door would in all probability include, accidentally from the villain's point of view, cutting the wire or foil. Given that this device is connected into the alarm system, an alarm signal would be triggered as it was cut through. After assembly of the wire to the door, it is customary to cover it on the inside of the door with plywood or hardboard to preserve appearances and to give physical protection against damage. Some suppliers of security doors build in this device during manufacture, with connections brought out via the hinges, hinge bolts or special contact blocks. These tend to be more expensive but less vulnerable to damage than loops of cable between door frame and door.

Provided that precautions are taken against corrosion damage, concealed wire and foil in doors is acceptably free from false alarm problems. Design details should be included in the Installation Code of Practice prepared by the security firm concerned.

Door contacts

As security locks improve in quality, in resistance to picking and in having an adequate length of dead bolt when thrown, say of 20 mm or more, the chances of a villain overcoming a lock are falling and the value of door contacts is correspondingly reduced. However, apart from installation costs and wiring, door contacts are inexpensive to incorporate in a perimeter door system, and should someone forget to lock a perimeter door, for instance, a door contact can ensure that the alarm is triggered if a villain takes advantage of that forgetfulness.

The range of types of contact is legion and is covered in some detail in Chapter 13, but it is likely that the basic concept of a reed switch fitted in the door frame, with an activating magnet fitted in the top of the door, will outlive other concepts of door contact.

Vibration sensors

An alternative to concealed wiring (page 37) to detect an attack on a perimeter door is a device made sensitive to vibration. Clearly an outer door using this principle would be very prone to false alarm were it not for precautions built into them to discriminate against wind rattling, a legitimate 'knock on the door' and mischievous banging by children to see what happens. These devices, on their own or in combination with door contacts, can take effective care of perimeter doors and the principles and practice involved are enlarged upon in Chapter 13.

Having reached this stage in the survey, it is time to ask the escort:

Question 4. We now know what we are going to do about the doors: what windows and other apertures are there in the perimeter walls that could be used by a villain to get into the building?

As you go round looking at windows, remember that your escort is there to help you, but not to do the job for you. If you miss an aperture and a villain subsequently uses it to cause a loss, you are unlikely to succeed in defending your company against a charge of negligence by saying that your escort didn't tell you about it. What makes the problem worse for surveyors is this: During a market research exercise I asked a very experienced man responsible for many buildings whether there were any gaps in the range of detection devices available for buildings. Unhesitatingly he said yes, there was nothing really suitable for windows. By suitable he meant unobtrusive, reliable, versatile, easy to install, free from false alarms, yet certain to detect intrusion. Oddly enough, he didn't mention price: there was nothing really suitable at any price.

Tube and wiring

To the casual observer this device looks like iron bar physical protection for the windows, but normally fitted inside rather than outside the glass. To gain entry after breaking the glass undetected, the intruder has to force apart the 'bars', which are in fact tubes containing concealed wires which snap when the bars are bent, thus triggering an alarm signal. While offering something of a deterrent, tube and wiring for windows is simple and effective. On aesthetic ground, and on others such as difficulties in window cleaning, the fact that they have to be tailor made and incur site labour costs, they are increasingly being phased out in favour of vibration devices referred to above (page 37) or breaking glass detectors.

Breaking glass detectors

As explained in more detail in Part 2, breaking glass detectors are a form of microphonic sensor, using piezoelectric material to convert the mechanical vibration of breaking glass into the electrical energy needed to trigger an alarm. The sharp cracking noise made as glass is broken contains vibration frequencies well into the ultrasonic region in addition to the obviously audible sounds, and in a well-designed breaking glass sensor only the energy at ultrasonic frequencies is used for triggering the alarm. In this way false alarms from almost all normal sounds, including banging on the glass, are ignored by the alarm triggering circuit.

   Thus, the piezoelectric breaking glass device goes a long way towards meeting the requirements for an effective window intruder detector, particularly for windows having only a single pane of glass. The limitation comes when the window is formed from multiple panes of glass, because the required ultrasonic frequencies are not transmitted very effectively from the pane of glass being broken, through the window frame, to the sensor. It is hardly practicable to fit a sensor to each pane, so other ways have to be used – but what? Try this as a discussion point at the end of the chapter. In practice it is wise for the surveyor to discuss window problems with the engineering staff before going firm on any one method.

Alternatives for window detection

Don't be surprised if the engineer says, 'Forget it'. Windows are an engineering headache, and do not provide much resistance to an intruder determined to get in. If the intruder is coming in anyway, it can be better from the system point of view to detect him as soon as possible after he gets into the building, rather than when he is trying to get in. If leaving detection that late seems too risky from the overall security point of view, the surveyor can insist on true perimeter detection at the window, and then be prepared for it to cost more, owing to the high labour content.

A simple compromise might be reached by using large pressure-sensitive mats inside the windows, to trigger the alarm when stepped on, or by using a point-to-point infrared beam across the window, but both can be evaded by a skilled intruder.

The most effective alternative to detection at the window itself is likely to be space detection, using equipment which triggers the alarm by sensing the motion of the intruder as he moves into the risk area. An infrared curtain detector might meet this need.

Which takes us to the next question to be put to the escort.

Question 5. What are the areas and items at risk within the building?

The risk can be anything from an open warehouse area to a single safe. The common factor almost always is that the risk is surrounded by space and the intruder has to move within that space to reach his objective, so, again, space detection is likely to be the most appropriate and we must look at ways of using it.

Space detection methods

The principal methods are ultrasonic Doppler, microwave radar and passive infrared. These are described in some detail in Chapters 15, 16 and 17, respectively, which need to be studied by surveyors to help them to decide on the method best suited to each problem.

Taking the window problem, for instance, passive infrared detection is the likely winner, as the energy does not leak through glass. Thus, the choice is based, not on detection capability, but on avoiding false alarms due to possible detection of legitimate movement outside the window. For larger areas, and for areas away from windows, microwave detection can be a better choice, since fewer units are needed and the longer wavelength reduces false alarm risks. The passive infrared detectors tend to come into their own in compact areas, as in most domestic risks, and in areas that an intruder must pass through to reach his objective, such as corridors. But these brief summaries are unfair to all the devices, and it really is necessary to refer to Part 2 of the book for a fuller understanding of applications.

We still haven't looked at the roof of the building, so, referring again to the escort, we ask:

Question 6. What access is there to and from the roof, inside and outside, including fire escapes, skylights, watertanks and ventilation ducts?

If extensive use has been made elsewhere of space detection, it may be unnecessary to do anything about the roof, but each job is different and only by going there and visualising what a villain might do can you decide how and where to detect him.

System control equipment

We established at the beginning of this chapter that the building being surveyed was used on a lock-up-and-leave basis and that the security system was needed when the building was empty.

Police experience shows conclusively that the greatest incidence of false alarms occurs at the opening and closing times of buildings. In other words, it occurs when the intruder alarm system is being switched on or off.

No matter how well the survey for intruder detection has been done, it will be worthless if the police refuse to respond to an alarm call because of an excessively high false alarm record. At least equal attention must therefore be given to the design of the system control equipment to minimise the risks of human errors.

Equipment design principles are covered in Chapter 24, but the surveyor is more concerned with system design, and this leads him to ask:

Question 7.   What is the present locking and opening procedure for the premises?

The chances are that the escort does not know in sufficient detail, and it will be necessary to talk to the one or two individuals who actually do the job.

Co-operation comes more naturally and the risk of error is minimised if the alarm system control routine to be introduced can be made to fit in with the user's normal procedure. But nothing should interfere with necessary discipline in usage, and allowance should be made for training of keyholders and users.

For engineering and maintenance purposes the various detection devices are likely to be connected back to a centralised multiway control box, and for operational reasons this may be located near the final door of exit. It is typical for the final switch on the control panel to start a time delay unit, to give the keyholder time to leave and lock up the building, but all too often a telephone rings, or the keyholder goes back to his office to collect something he had forgotten, and the time delay expires before he has got back to the final exit door. The now-active detection system thinks he is an intruder, and so another false alarm is recorded.

There seems to be only one satisfactory answer to this problem of human nature, and this is examined in Chapter 24. Briefly, it involves using the central control panel to switch on everything, including local audible alarms, but excluding the remote alarm signalling circuit to the police or central station. The remote signalling circuit is connected through only when the final door of exit has been locked closed from the outside. As we shall see, there are engineering problems, but these are less than the human ones.

Tampering

Remembering the dual question, 'Will it work? Can it be made not to work?', there is another point to clear up with the escort.

Question 8.   What will be the attitude of personnel in and around the building to the introduction of an effective intruder alarm system?

Given obviously co-operative personnel, no special precautions are needed other than the normal night-time antitamper circuits designed to trigger if an intruder thinks he has found a way of putting the alarm system out of action. But there are ways a malicious employee can tamper with an alarm system. Without a tamper-sensing circuit operating during working hours the customer is unaware of tampering until he finds that he cannot set the alarm correctly at the end of the day.

Typically, it is the service department of an alarm company that bears the brunt of this problem, concentrated into an hour or so at their customers' closing times.

If there is a prospect of tampering, the surveyor should make sure that his engineering department design the system accordingly with a working-hour tamper alarm fitted in an appropriate office or location for attention by nominated individuals such as the security officer or the works manager.

Alarm signalling

The survey we have been following has served as an introduction to the various elements of an indoor intruder detection system. The last, or first, link in the chain, depending upon your viewpoint, is the signalling system between the installation in the building and those who have to deal with intruders – normally the police.

The principles involved in this aspect are examined in Chapter 8 and application of the principles in Chapter 24. A good surveyor will be well briefed on signalling, on his company's policy and on police policy so that he can advise his escort on what is involved.

Preparation of proposals

By now the surveyor and the escort should be clear, and agreed, upon the overall security system design. But the job doesn't end there. It has to be priced, and it is customary to write down what the customer would get for that price. In other words, the surveyor has to write a proposal.

The surveyor must beware not to write a report. A report covers a job that has been done (and so it has so far as the surveyor is concerned), but the security system itself is still to come. 'Proposal' seems to be the right word to use, and any surveyor worth his salt will want to have his proposals accepted.

Who then has to be persuaded?

Question 9 for the escort might well be, 'Who makes the decision to go ahead?'

Ask to see him now, while it is all still fresh in your mind. Surprisingly often, directors and decision makers will make time to see a security man, and this gives the surveyor the opportunity to ask what the security system is expected to do. Again perhaps surprisingly, top people are often very frank with the likes of us on security matters.

A few minutes of discussion, afterwards verified with the escort, can be invaluable in establishing just what the surveyor should write in the proposals under the heading of 'Purpose' or 'Objective'.

Whether the decision maker reads the proposals himself or has the salient points picked out for him by, say, the escort, it helps them to find their way around the proposals quickly if frequent use is made of headings such as 'Objective', 'Method' and 'Conclusions'. Sub-headings help too, such as 'Vulnerable points', 'Opening and closing procedure' and 'Actions in the event of an alarm'. Each should be discussed from the point of view of the user, the customer, and not from that of the surveyor.

Even so, the same words mean different things to different people, and if at all possible, the surveyor should contact the escort shortly after despatch of the proposals to check that the proposals have been well received and are free from ambiguities.

Chapter 29 on Presentation of Information may be helpful for further reading.

Acceptance of proposals

This book is as much for the customer and user as it is for the supplier of security systems. So let us now look at the proposals as users, and see whether we can accept them or recommend their acceptance. Even if you have a long-standing understanding with one supplier, it is important to have an alternative proposal from another supplier of similar reputation as a cross-check on techniques and prices, without implying disloyalty to a firm that has served you well. Suitable possible suppliers may be suggested by the insurance company if they are involved, and it is worth consulting the police, as it is not their practice to suggest only one supplier.

Instinctively we look at and compare the prices quoted with the various proposals. Difficult though it is, it is crucial to resist drawing conclusions from a first glance at the figures. If this were a book on selling, it would have quite a lot to say about firms that quote prominently a low price, and leave the customer to ferret out from within the text the cost of necessary 'extras' needed to complete the system.

Such firms excuse themselves by saying that they lose out with customers who do decide which firm to go to from the first quick glance at prices. So, don't encourage them: the only way to be fair to yourself and to suppliers who make a genuine attempt to show you the real cost is to tabulate the features of the systems and to read the proposals and fine print in detail, extracting and tabulating all the figures, provisos and exclusions.

Tax law puts different emphasis on capital expenditure and on revenue expenditure, and if maintenance is to be included in the contract, the quotation may cover both capital and rental charges. Suppliers can put the emphasis either way to suit themselves, but to be clear in your own mind on likely total costs, it is worth assessing what the proposed systems would cost over a given period, say five or ten years.

Total cost is a concept I favour, but it is your accountant who has to be satisfied, and he will advise which proposal suits him best. If that does not match your choice of security system, the preferred security system supplier may be prepared to listen to the accountant – it is worth a try.

Obligations in implementation

So much legislation has come into being in recent years in many countries that few people can be unaware that the customer or user has legal rights regarding fitness for purpose of products and services he purchases. If he buys a mowing machine, or she buys a sewing machine, the user knows what to expect of the machine, and is soon in no doubt whether it does or does not work as it should.

It is not so straightforward with security systems. There is the ever-present hostile environment created by the villain, who has a vested interest in seeing that the system does not work; there is the communication link between the system and the police which is outside the control of both parties; and the police may be too busy elsewhere to react in time to an alarm call.

The elements of divided responsibility and vested interest make it even more important in security work than for consumer goods to avoid making claims that cannot be honoured with certainty. It is no use saying that the system will call the police – it can only do that with the co-operation of other services, and if the police do not react as expected, the burden of proof that the failure did not lie with the security system itself is a heavy one.

Customers tend to demur when statements such as 'The security system *is designed* to call the police' are used in proposals, but with good faith the statement can be true. No amount of good faith can make the statement 'It will call the police' true.

This is yet another example of the fact that security is part of the risk business. In order to reduce the risks to the customer, however, Government-inspired regulatory bodies exist to produce and to implement Standards and Codes of Practice, which set out the obligations of security organisations in the implementation of security systems.

Further illustrations of the obligations involved appear from time to time in later chapters, but first we shall be looking at the much newer subject of movement detection outdoors.

Discussion points

It is no bad thing for anyone in the security business to spend some time looking within themselves to find out what they really believe. Earlier in this chapter it was questioned whether it was right or not to use fear as an inducement to have security equipment installed. If you accept that the prevention of crime is our function, it can be said that the use of fear is justified if it helps to achieve our objective. If working on the emotions is not justified, what other methods can be used?

Another discussion point arose when we found how difficult it is to detect intrusion through a window. Have those who have worked on the problem so far overlooked some very elementary fact which could lead to a solution? Can you invent a better way?

6 Movement detection outdoors

Why outdoors?

In order to set the scene for this chapter, let us first consider the need, the environment and the technical conflict involved in the outdoor detection of intruders.

Regarding the need; we can take one or two examples of outdoor detection, starting, say, with fish farming. An unlikely subject? Maybe; but think of the man with tank after tank of trout, at various stages of growth. The market for trout is good enough to attract villains, and they do not take the small ones. No amount of production engineering will speed up the growth rate of the remaining fish sufficiently to compensate for the loss of the big ones; all that our fish farmer can do is wait. The point of this example is two fold – it illustrates the trend of the intruder problem to outdoor risks and it illustrates the problem of indirect loss, or, more precisely, of consequential loss as well as direct loss, and consequential loss is much less adequately covered by insurance.

At the other end of the scale are the national risks arising from people trained in subversion, intimidation and sabotage, as referred to in Chapter 3. Whether the target is a fish farm or a military site, the sooner we can detect the approach of the attack, outdoors, the greater our prospects of defeating it. If in spite of our efforts at deterrence an attack does develop, then physical protection and delay are given by fences, walls, window bars and the like, and warning that an attack is developing is provided by the outdoor detection system.

We say 'the outdoor detection system' as if it were customary and natural for it to be provided, and as if the selection of an outdoor detection system were easy and obvious. Neither assumption is true.

The environment is hostile, to friend and foe alike. Out of doors the weather itself is hostile, but security equipment must operate come what may, be it rain, snow, hail, fog, lightning, heatwave or ice. And, of course, the intruder is hostile, and he will evade or tamper with anything he can to immobilise the detection system before setting about his objective.

All of which brings us to the technical conflict – that is, the need for, and extreme difficulty of, avoiding false alarms outdoors while retaining certainty of detection. Outdoor detection systems are aids to security manpower, and if the credibility of a system is impaired by false alarms, human nature tends to take over from discipline and a real alert may well be ignored in the belief that it is just another false alarm.

It is necessary to note here the two operational elements in security – the active part played by manpower, and the passive part played by the physical and technical aids to that manpower. Emphasis on gaining time tends to concentrate our attention on the perimeter of the site but economic balance suggests that we do not spend all our resources just there. The law of diminishing returns applies to this work as to any other, and a sufficiently impregnable perimeter will lead a determined intruder to consider methods of evasion. Tunnelling is traditional, and the helicopter and the cherry-picker are no longer futuristic perimeter evasion aids – both have been used successfully. Resources should be properly balanced to give second and possibly third lines of defence for individual vulnerable points and areas within the perimeter.

Physical properties of an intruder

With that introduction to the needs of the user, we can move on to consider how electronic techniques can be developed and applied as aids to security manpower. To understand and to select the most appropriate intruder detector for each outdoor application, it is necessary to be aware of the physical properties of an intruder and of which of these properties can be detected with sufficient certainty to raise an alarm. Coupled with this has to be the ability of the device to discriminate between the intruder and everything else if false alarms are to be avoided or minimised.

It is stating the obvious to show how little there is to distinguish a man from an animal. He has weight; he breathes, exhaling carbon dioxide gas; he is optically opaque; he has a distinctive shape; he moves on two legs and sometimes four, with rhythmic motion. He can climb; he can see, speak and hear. He can think and reason. He can use tools, to dig and to cut. He can jump and crawl, and whether he likes it or not, he radiates body heat. He is made mostly of water, he has nerve, and nerves, purpose and fear. In the discussion section you can add to the list but meanwhile we will go on to see what can be done with the properties mentioned so far.

Initially, selection is intuitive, and we tend to concentrate on properties for which sensors or transducers have already been developed for detection indoors or for other applications.

Infrared beams

Take, for example, the infrared beam, so widely used for all manner of jobs. Here the opaque nature and motion of the intruder interrupts reception of a beam of infrared energy transmitted towards a receiver. This method became discredited for a while for outdoor use because of attenuation and eventual extinction of the beam in fog and heavy rain, causing false alarms and the risk of successful penetration.

More recently this method is enjoying renewed interest through the use of higher power and more reliable sources of infrared energy, and even if cut-off does still occur in thick fog, a good question to ask is, 'Does it matter?' Statistically, in the UK at least, thick fog occurs but rarely, and what are the chances of the intruder making his attack to coincide with such foggy conditions, and if he does, would not the inner lines of defence advocated above be sufficient to detect him in time?

If it does matter we have to find a better way.

Using longer wavelengths

The beam technique is attractive, and rather than look for another principle, we should look first at the application of the principle. Although the method of use may be unsuitable, the principle itself may still be correct.

The real restriction with the infrared beam is that the wavelength is too short, and the excessive loss of beam strength in rain and fog happens because the wavelength of infrared energy is too close in size to the water droplet diameter of fog and rain, which causes absorption and scattering of energy.

To avoid excessive attenuation of energy in rain and fog, we have to find a device which will transmit and receive a beam at a longer wavelength, which will not be lost en route. But the wavelength must not be so long that we lose control of the shape of the beam – it could become too wide, as described in Chapters 4 and 16.

The microwave fence

Holding to the idea of using the same principle as that used in the infrared beam, we find what we want by using microwave energy. In spite of the name, the wavelength is much longer than for infrared light. First generally known for radar applications and then in point-to-point communication as a replacement for cables, eventually for quite different reasons, microwaves came into the home, for use in cooking.

The configuration closest to what we want as an all-weather alternative to the infrared beam is the microwave beacon used for point-to-point communication, which, when developed for intruder detection, became known as the microwave fence. The three main versions of the microwave fence are described in Chapter 20. The first two versions operate on the same principle as the infrared beam, triggering when the beam is interrupted by the intruder, but the third version relies upon the intruder reflecting rather than blocking the microwave energy.

There is no need to get bogged down on this problem of wavelength. An understanding of it helps, and to the designer it is all-important. But the systems man can get by without it for the time being.

Alternative properties of the intruder

Even if the microwave fence is the answer to problems of infrared beams outdoors, microwaves are not the answer where space is severely limited, where the direction of the perimeter changes frequently or where the ground undulates appreciably. This is simply because microwaves, like light, travel in straight lines over the distances we are concerned with, and it is too costly to use a pair of transmitters and receivers for each change of direction.

Our control of the situation ceases outside the site perimeter, where legitimate presence and movement of people, animals, vehicles and vegetation have to be accepted. So, by being forced outwards by space limitations and inwards by the legitimate boundary, we may have to concentrate upon the wall or fence of the perimeter, where the risk of false alarms seems to be at its highest.

Alternative properties of the intruder that are most likely to be useful are weight, and change of velocity of movement, which together can induce vibration. The vibration can be in the wall or fence, in the ground adjacent to them, or both.

Various devices have been created or adapted to sense these vibrations for intruder detection.

Vibration sensors

The original geophone and vibration contact had to give way when the false alarm problem was better understood. This understanding led to the realisation that the vibration sensor itself could be designed to reject a significant range of false alarm situations, rather than relying overmuch on subsequent electronic signal processing. This is just as well, for, as we have seen, perimeter detection has to work in the worst of environments and everything that electronic signal processing can do is needed as well, to help create the complete system.

The *Geophone* has effectively been redesigned to cater for the larger amplitude signals encountered in security relative to those in its original application in seismic surveys. The geophone can discriminate between head-on and sideways impact, but for other ways of inherent false alarm rejection the trend has been towards devices which are deliberately sensitive to acceleration or *G* forces.

Inertial switches. Nature is with us here, in that many causes of false alarm are at low frequencies, to which *G* devices are not sensitive, while intrusion tends to generate high frequencies, to which *G* devices are particularly sensitive.

Several forms of sensor have been evolved, in which an inertial mass has been chosen to give the required rejection characteristic, sometimes combined with artificial gravity – a magnet. Active devices include piezo-electric sensing using individual devices, or distributed sensing, as in the electret cable. Passive devices include the liquid mercury ball; and others have a solid metallic ball or roller operating on the so-called inertia switch principle, which strictly is as we want, sensitive to *G* or acceleration.

In spite of all the problems of detection at the outer wall or fence of a site, continued development has produced a range of systems using the acceleration principle applied in the various ways mentioned above. But when the properties of the intruder are compared with those of an animal and other objects causing false alarms, it has to be accepted that there is a limit to what can be done, and putting sensors on a structure dividing the risk from the public at large is asking for trouble from false alarms.

Improvements in performance are being made all the time, but, allowing for that, I would say from my experience that security system people should resist using sensors on outer perimeter walls and fences. If pressures are such that they have to be used, then any alarm from them should be treated as early warning – amber, and not as a full red alert. In this way the integrity and respect for the system can be maintained in the minds of the site uniformed operating personnel.

The role and attitude of security personnel

In the aspects of detection of villains out of doors that we have looked at so far, there are more problems than answers. Difficult though it is, we must move forward persistently to make sure that our electronic aids are aids and not a nuisance to the personnel responsible for protecting the site.

Once one gets down to sorting out the problems of any outdoor site, the first thing to do is to find out in principle where the operational security personnel are to be located.

Categories of security personnel

As established terminology is none too good, it may not be clear who does what. However, the following categories and duties of security personnel can be distinguished:

(1) Those stationed within a security control room to monitor the alarm indicators and to inform others of any incident requiring their attention. It would be contrary to good organisation to allow control room personnel to mix these duties with any other duties, and, in particular, they should not themselves go off to investigate an incident while still on control room duty.
(2) Patrols are those who show a presence anywhere within the site but should not be required to go outside it. Their duty is to report situations and incidents to the control room, but they are non-combatants. They are typically of the same status as control room personnel, with whom they may exchange duties on a timed rota basis to reduce fatigue.
(3) The above categories may be on the payroll of the owners of the site or they may be employed via an independent agency. Questions of economics, loyalty and expediency on the source of manpower are relevant, but need not concern us here.
(4) The civil police are normally the ones called to a site by the control room personnel to deal physically with an incident, ranging from investigation, arrest of intruders on site or elsewhere, through to court proceedings. The civil police may or may not be armed.

(5) In exceptional risks, the reaction forces may be armed military personnel equipped to suit the risk.

Response to alarm signals

Bearing these categories in mind, we can consider again the implications of an alarm signal being received in the control room from sensors on an outer perimeter fence. What are the operational orders given to the control room officer? – Or what should they be?

If his orders are to treat all alarms as real, then he would call for immediate outside help from the civil police or for military help. We saw earlier that the risks of false alarms from outer perimeter fences are high, and it does not take many such calls for the detection system to lose the confidence of security personnel both on and off the site.

Ought not the orders therefore be to initiate a local investigation first, by sending the patrol officer to find out the cause of the alarm? As statistically he is likely to find the cause to be false, not caused by an intruder, he can report back to the control room officer and so save a fruitless call to the civil police and, at least to some extent, preserve the reputation of the detection system.

But supposing the patrol officer encounters an intruder or an organised raiding party? He will be equipped to signal confirmation of a real alert to the control room, but his own survival is very much at risk. Whether the patrol officer should be armed is not the concern of this book, but reducing the need for arms is.

It would seem, therefore, that neither treating all alarms as real nor requiring a patrol officer to investigate is an acceptable procedure in itself.

Let us be clear: I am not blaming the equipment or the personnel, but I am pointing out that through perhaps inadequate appreciation of the role of manpower in relation to their electronic aids the use of detection equipment on the outer perimeter of a site can lead to the seemingly unanswerable dilemma outlined above, and can bring the equipment into disrepute.

However, given attention to the basic principles outlined below, outdoor detection can be satisfactory from all the various points of view.

No man's land

The above arguments should be enough to enable the surveyor and system designer to persuade the customer not to have detection devices on the outer perimeter wall or fence. If the customer replies that there is no room for it anywhere else, a closer study usually shows that room can be made for it if the will is there. And the site will probably look the tidier for the effort.

One of the detection systems needing barely a foot of space within the perimeter, yet avoiding having any sensors on the outer perimeter itself, uses columns of infrared rays. If evasion risk or performance in fog discourages this concept, the next step is the fence within a fence, needing up to a metre of ground between the two. The outer fence would be the existing boundary marker and physical barrier to the public at large, with no detection devices fitted, but the inner fence would normally be an addition, and would be fitted with vibration-sensing devices as described above or others selected for the particular environment and risk.

Microwave fence and radar sensors

In situations where the no man's land between the fences can be wider still, up to, say, 15 metres wide, and where risk of evasion of fence sensors is considered significant, one of the various types of microwave fence described in Chapter 20 should be used, given, of course, that the ground is sufficiently level and the perimeter sufficiently straight. If activity on the inner perimeter can be controlled, it may be unnecessary when using a microwave fence to have a physical fence to mark the inner boundary of the no man's land.

Where there is an abrupt change of ground level in the route of a microwave fence, continuity can be maintained by having one microwave fence at each of the two levels and filling in the sloping gap with an outdoor microwave radar, as covered in Chapter 16.

There are also body-heat-sensing passive infrared detectors that can be used out of doors in place of radar to fill the gap between two microwave fences on abrupt changes in ground level.

Electret cable and E field fence sensors

For the higher levels of risk it becomes unwise to rely upon any one method of detection at the site perimeter. Given that there is an outer perimeter wall or fence with no sensing equipment, and then no man's land with microwave fence detection, an inner perimeter fence equipped with any of the fence detection systems mentioned above, or with electret cables or E field fence sensors, would provide the back-up detection system.

Electret cables and E field systems, described in Part 2, are of particular value as primary or secondary protection in undulating ground or where the site boundary changes direction frequently and at relatively short intervals.

Even systems having a no man's land and back-up detection facilities are still prone to false alarm from birds, rabbits, weather and vegetation, but their characteristics are sufficiently different from those of human beings and the larger animals to give electronic designers a chance of filtering out most of the false alarms in the signal-processing circuits. But 'most' is all that can be claimed, and a highly favoured further false alarm filter is television surveillance.

Television surveillance

This technique, usually called CCTV (closed-circuit television), is covered fairly fully in Chapter 7. Let it be said immediately that it is no good asking the control room officer to watch a row of TV monitors and expect him to respond reliably to an incident. But it is possible to have monitor screens coupled to the perimeter detection system in such a way that whenever a sensor approaches or reaches alarm trigger level, the sensor switches the picture from a camera surveying that area on to a monitor, with an audible alert to attract the officer's attention to that screen. With visual confirmation from what he sees on the screen, the security officer can act with confidence and can convert an amber early warning to a full alert, or confirm the alarm as false. In a suitably designed system he would have the option of overriding the automatic monitor switching to allow him to have one or more cameras switched through to monitors until he could confirm that all really was quiet.

However, incidents will not wait for the security officer to come back to his control panel if he is away making tea. CCTV is a waste of money and a source of dangerous false confidence if management do not make provision for two security officers always to be on duty together in a control centre. In all well-considered Codes of Practice on control room operation, this level of manning is a firm requirement, as is also the corresponding discipline to ensure that one of the duty officers is always in position to monitor the TV screens when activated by an intrusion sensor.

Therefore, the message for outdoor perimeter detection is get some no man's land, and have no detection devices on the fence dividing the protected site from the public at large.

Now we shall look at the protection of areas within the perimeter.

Vulnerable areas and points

It was emphasised on page 46 and subsequently that we should not spend all we have on the outer perimeter detection system. Adequate resources must be reserved for allocation to inner lines of defence, covering the approaches to and the immediate vicinity of vulnerable areas and points. A useful word for this is 'zoning'.

The proportions involved are a matter for judgment, taking into account the nature of the risk, the existing and proposed levels of physical protection, the deterrence effect, evasion and the times involved in penetrating to the risks themselves in the event of a determined attack.

Although the problems are easier within the perimeter, system design needs to take account of the legitimate activities of personnel within the site, such as grass cutting, storage and movement of materials, contact with equipment that can cause misalignment and new building operations.

Once decisions have been made on what measures are to be taken, continued middle and lower management co-operation is necessary to maintain the levels of awareness and tightened discipline needed with improved site security.

Presentation of information

The design of the system for use within the perimeter can be based on a selection from any of the detection devices already mentioned, and covered in more detail in Part 2, ranging from vibration sensors through to overall CCTV surveillance.

Information to control room officers

The larger the site the more important it is to realise the need to make it instinctively easy for the security officers to know just where on the site an incident is occurring. They will have their own jargon and it is worth finding out what they call various locations. It is no use referring to the north-east corner if that part is universally known as the pit.

This feeling of instinctive orientation is aided in the control room by the use of mimic or graphic displays. These either consist of a wall- or console-mounted chart of the site, with indicator lights and associated controls, or the site and alarm information can be presented on a video monitor. With the monitor the operator can have the choice of displaying just one zone of the site, or the whole site.

Information for management

When it comes to preparing system proposals, it is as well to be aware that management – the customer – and their representative – your escort – are likely to be less familiar with outdoor detection than they may be with indoor detection. Even greater attention is therefore needed to ensure by frequent discussion that your escort is party to your proposals. This discussion should extend to offering a draft of the proposals for his comments before you submit the proposals formally.

Another factor to work into proposals is the need for trials of the system after installation and before hand-over. One cannot think of every eventuality, and you don't want the customer to keep saying, 'Oh, you didn't tell me that'.

One example may illustrate the point. During trials of an outdoor installation on a waterway, the CCTV lights came on from time to time during the night for no apparent reason. The reason was real enough. A nocturnal boatman was seen to come over the low boundary fence between the site and the adjacent boat club in order to trigger the detection system just within the boundary to bring on the CCTV lights so that he could see more clearly what he was doing. A higher fence between the site and the boat club put a stop to that, but the need had not been appreciated until the system was undergoing its trials.

Discussion points

What physical properties of an intruder do you consider have been neglected by the security industry in the design of intruder detection devices? What properties were omitted from the list considered on page 46.

What is a 'moderate' false alarm rate? Is a moderate false alarm rate beneficial in helping to keep control room security personnel alert?

7 Surveillance

The eye as a source of information

Seeing is believing. How true is that? To test a cliché of this kind it is handy to consider the opposite effect. If an alarm signal sounds and a light comes up on a mimic panel, what can the security control room officer learn from that? If he cannot see what caused the alarm to trigger, he has to obey his instructions blindly, which probably means that he raises a full alert for the reaction forces, not knowing whether the alarm was true or false. It only takes a succession of false alarms for his confidence in the system to be shaken, and when that happens, as we found on page 46, human nature tends to take over from discipline, with the consequent risk that he will, some time, ignore a real alarm in the belief that it is yet another false alarm.

If only he could see what was going on. If he could see with his own eyes through the control room window someone breaking through the fence, he would know the alarm was true. Seeing is more than believing – it is knowing.

Extending the range of the human eye

The idea that the control room operator really could see the perimeter fence, all of it, all round the site conjures up visions of a tower in the middle of a concentration camp. Even then, the guard has to have eyes at least in the back of his head to be sure that he is seeing all the perimeter all the time. Failing that, he needs help, not just from a colleague facing the other way but electronic help, in the form of television cameras and monitor screens. Given a comprehensive TV system, he no longer needs the tower, the control room can be at or below ground level, and if I were he, he would be only too glad to be out of the way of any villain ready to take a potshot at an elevated observation post.

As presented above, the reader can be excused for gaining the mental impression of the need for rows and stacks of TV display monitors, each connected to cameras placed round the perimeter fence. He can be excused because that is the impression he is meant to have. This is another technique for clearing one's mind – to reduce a situation to absurdity. It is absurd in my view to think that a large collection of monitors in a control room is an acceptable way of presenting information from the cameras. This view is, I know, controversial but I would like you to think it through for yourself.

It is no use condemning a concept unless one can offer reasons and alternatives. For reasons we have to look no further than human nature again, and a little at physiology. And there are at least two good alternatives.

Boredom

One aspect of the problem is boredom. If each camera is connected to a monitor, the picture presented on each monitor will hardly change week in and week out. Laboratory tests and extended field experience show that there is a duration beyond which a good security officer cannot concentrate on a monitor screen and react reliably to a significant change in the scene, such as the entry of an intruder. Even after only a short spell of duty there is an unacceptable risk that through no fault of his own he will fail to see the intruder. An uninteresting domestic TV programme often sends viewers to sleep. Because of the absence of the element of duty, the domestic example is not strictly a fair comparison, but the element of boredom is the same in both cases and the resulting inattention can be the same.

Physiological aspects

The physiological aspect of the problem is concerned with a curious property of our eyes, which shows up in two ways. When we look straight at a domestic TV screen or at a security TV monitor screen, we see a complete picture. But that is an optical illusion. All that is on the screen at any one instant is a very small dot of light, and the rest of the screen is blank, unilluminated. But electronic circuits within the TV system cause the spot to move very rapidly from left to right and from top to bottom of the screen in a series of horizontal lines, and this is repeated 25 or more times a second. The camera scans the scene in the same way, and the brightness of the light in the dot varies from nothing to very bright, to match what the camera sees. The fact that we don't see the dot itself and we do see the complete picture is called persistence of vision. The effect is the same, but created slightly differently in a ciné camera and projector, where the picture is changed about 24 times a second.

But not all of the eye works in this way to enable persistence of vision to give us the illusion of a continuous complete picture. Only the centre part of the eye does this, and hence the need to look directly at the screen.

If we do not look directly at the screen, the picture seems to flicker, because 'out of the corner of one's eye' there is practically no persistence of vision. This is probably a relic of prehistory, when it was necessary for man to be instantly aware of a predator preparing to attack, but it is just as useful in today's traffic conditions.

You can try this next time you watch TV or a monitor screen. When there is no significant movement in the picture, as with a TV caption or a normal monitor scene, look straight at the screen for a few seconds and then shift your eyes to look at an object in the room some 30° to right or left of the screen. As you look at the object the picture on the screen will be seen out of the corner of your eye. But instead of being steady it will flicker.

Going back now into the security control room, still with its many monitor screens, it is often argued, 'But surely when the picture does change, as with an intruder entering the scene, the security officer will notice it, even if he is looking at another monitor?' An understanding of 'the corner of the eye flicker' shows how risky it is to rely upon the operator seeing a scene change. Partly from boredom, but also because the flicker is there each time he moves his eyes, he gets used to it to the extent that a scene change is just another flicker. As was mentioned earlier, without being negligent in any way he just does not notice it.

Fur further reading you may care to study the parts of the eye's retina called rods and cones.

Video motion detection

Those are my reasons for discouraging multiple monitor installations. For solutions let us look first at a technique which is best expressed by the word 'move-alarm'.

'Video motion detection', to give it its functional name, is quite simple in concept, quite complex in design and circuitry, and quite easy in application. Here we need concern ourselves only with the easy bits.

In concept video motion detection acknowledges the problems of human nature and physiology described above, and says in effect, 'Let us do something to attract the attention of the operator to a monitor showing an incident'. In application a move-alarm system can look much the same as any other multiple camera and multiple monitor TV surveillance system. The difference is that circuitry is introduced to sense when the picture changes. The picture from a camera looking at a fence cannot change much unless an animal or a person appears on the scene. Once that happens, the circuitry can be arranged to sound an alarm immediately to attract the operator's attention and at the same time to light up an indicator lamp on the monitor showing the intrusion, telling the operator to 'look at me'.

The operator really would have to be negligent to ignore these warnings, so operationally and in concept the move-alarm principle is sound. Equipment offering this facility is described in Chapter 22.

Reducing the number of monitors

In the move-alarm concept it is not necessary to watch any of the monitors until something happens. That is ideal. If it is not necessary to watch them, isn't it rather a waste of monitors – why not use just one, with perhaps one or two spares, making two or three in all?

Originally the concept involved incorporating the movement-sensing electronics within or on the face of the *monitor*. As an alternative solution there is no technical reason why the electronics should not be incorporated in the *camera* circuitry. When any camera notices a scene change, it signals to an electronic switch 'show me' and the switch puts that camera through to one of the monitors.

With either method we are free to get rid of most of the monitors, leaving just the two or three necessary ones, which can readily be built in to a comprehensive security control console, where everything the operator needs is readily to hand and eye. Clearly the move-alarm concept is suitable only for normally static scenes. Applications involving active scenes are covered in Chapter 22.

Using other intrusion detection methods

A further alternative uses two of the ideas expressed above – that it is not necessary to watch a monitor until something happens and that no more than two or three monitors are really necessary. But for sensing a change in a scene this alternative is quite different.

On any installation incorporating intruder detection devices such as those mentioned in Chapters 5 and 6, cameras can be installed to survey the areas covered by the detection devices. It remains only to arrange that when a detection device triggers to raise an alarm, it also switches the camera associated with its area of coverage through to one of the two or three monitors in the control room.

Apart from the risk of breakdown, the possibility of a diversionary attack or of an attack centred between two adjacent sensors which may both trigger in a given incident makes the provision of two monitors necessary. A third monitor is sometimes needed to maintain a continuous though brief watch on a particular risk area, or to be associated with a sequential switch, which will automatically connect each camera in the system in turn to the monitor for five seconds or so as a routine check that each camera and lighting section is working correctly. Any increase on three monitors starts to take us back to the problems we have already discussed that arise from multiple monitor installations.

So far I have avoided calling these installations closed-circuit television, or CCTV. 'Closed-circuit' means connection between camera and monitor being by cable, to distinguish it from television broadcasting, where the link between camera and TV receiver is by VHF or UHF radio. But times have changed and domestic TV by cable link is established practice, just as a radio link can be used in site security where it would be difficult or expensive to provide a cable link.

In spite of these contradictions, the term 'CCTV' lives on, and so long as it is taken to mean private TV as distinct from public service TV, it is as good a term as any to distinguish it as a form of security surveillance.

It is necessary now to look from a systems point of view at some of the features and factors involved in the use of CCTV. As usual, the equipment itself will be covered in Part 2.

CCTV system concepts

CCTV is strongly indicated for locations where it is necessary to know, without doubt, what is going on, but where it is not practicable to be sure of having a man there to see for himself and report back. Please note: what *is* going on; not what *was* going on.

As we reasoned earlier, because a control room operator through no fault of his own cannot be relied upon to notice a change of scene just by watching the monitor screens, the CCTV system has to be linked with some form of movement detection device to attract the attention of the operator.

One of the areas of greatest uncertainty as to what is going on is the outer perimeter of a site, where there is also a high risk of false alarms. The combination of a perimeter movement detection system with a CCTV system probably gives the optimum solution to the uncertainty problem, provided that camera and siting selection are properly done in the system design.

Camera siting

The variables in camera siting include height from the ground, angle of view, distance from target area, direction of view, fixed or variable line of sight, availability of natural and artificial light, light levels, position of the Sun at all times of the year, relation to adjacent cameras, blind spots and overlaps, tampering, cable routes, ease of maintenance, and, as in all our exercises, cost. Unless someone will pay for it all, you are wasting your time.

With this as a pattern, you can prepare your own check list to work through when on site, and the check list itself needs to be tested against the overall objective of helping the control room operator to be sure he knows what is going on, so that he can take the right remedial action in time.

'In time' means that there is not much time for him to think; he must be able to assimilate almost instinctively what he sees on the screen. On the screen a man must look like a man, and a dog must look like a dog. And neither will happen if the image on the screen is too small. This gives the starting point for siting of cameras.

If you can, try using a camera and monitor and a volunteer target to get the feel for the size of image you could have confidence in, making allowances for an operator suddenly being alerted out of boredom. My own yardstick is to try to get the image of a man not less than 25 mm high for the target at the maximum distance from the camera. This leaves me with something of a bargaining margin for economics and siting difficulties. Would you go for less than that?

Camera selection

The above approach to camera siting leads in to camera selection. The first need in minimising uncertainty is good picture definition. Lighting, cables and monitors all play a part, but it is the camera quality that governs the overall performance. Lenses should no longer be a problem in TV cameras, although, as a general guide, the shorter the focal length, and the smaller the aperture, the better the chances of retaining a well-focused picture.

It is the picture target inside the camera, equivalent to the retina of the eye, that is the limiting factor, and more is said about this in Chapter 22. Camera electronics also should not be a problem, but the combined performance of electronics and camera target is revealed by the number of individual picture elements per line of scan (the overall resolution, equivalent to the number of dots per line width that go to make up a newspaper reproduction of a photograph).

In deciding what picture quality is good enough, the normal requirement for outdoor surveillance is discrimination between a man and everything else. In a banking hall raid, for instance, the alarm is real all right and the need is for maximum information to help in identifying the raiders.

Apart from definition, the other major factor in camera selection is a system decision on the lighting and camera target type combination. The sensitivity of the target in a monochrome camera to different colours of illumination varies from type to type, and the type used needs to match the lighting used, and vice versa. Similarly, for a colour camera the spectrum of any lighting used must include any colours that are required to be reproduced by the camera plus monitor plus recorder combination. For instance, tungsten filament lighting has very little blue content, and consequently, any blue items in the scene (a shirt say) may be reproduced as black.

However, an obvious variation is in situations where it is required that the intruder should not know that he is under observation, and the scene is to be lit with infrared light. This is invisible to the human eye, yet TV camera targets can be designed to be particularly sensitive to it.

Monitor selection

As for cameras, the design features leading to monitor selection are more matters for engineering than system decisions and are covered in Chapter 22. The principal factor concerning the systems man is the size of the monitor screen. There comes a point in reducing the size of a monitor screen when the size of the spot of moving light that builds up the picture, through persistence of vision, cannot be made smaller in the same proportion. This problem arises from a combination of limitations in focusing, halation and screen material. So, information that reaches the monitor is lost at the last moment of presentation behind the screen. These problems are almost completely overcome by making the screen bigger, and all the available information is presented on the screen.

If the screen is made bigger still, it is unusual for the spot size to be increased in the same proportion, so individual lines become distinctly visible, with space between them. In a sense, this is wasted space, and can distract a viewer from instinctive appreciation of the scene.

An additional factor is the distance between the viewer and the screen. Not a great deal of scientific investigation was done to find the optimum combination of these factors, but the widespread use of monitors for information transmission has led to a more thorough study of the requirements. Quite properly the most convenient viewer-to-screen distance was established and then the least tiring viewing angle was found. This gives the screen size and then the electronics were designed to give optimum definition for that screen size.

It seems that a more studied approach to the problem has produced much the same answer as intuitive selection, giving typical monitor screens having a diagonal size of between 9 and 12 inches.

Natural and artificial lighting

We really ought to be grateful for the Sun. We don't have to buy it, and it costs nothing to run. If we want to be able to see when it is not there, either directly with our eyes or indirectly with electronic aids, we have to provide artificial light, which can be costly to buy and very costly to run. It makes sense, therefore, to find ways of using as much light from the Sun as possible and to find ways of minimising the use of artificial light.

Optimum use of available light is achieved by using cameras fitted with the so-called low light level targets. The high sensitivity of these camera targets enables them to be used down to twilight conditions, after which artificial light is needed, but not so much as when 'normal' camera targets are used.

Low light level cameras are more expensive in first cost, so there has to be a trade-off between higher initial – or capital – costs and lower running – or revenue – costs. This trade-off can be assisted by so connecting the lighting system that it is only switched on when an intruder detection system says that it is needed, to illuminate the area concerned both as a deterrent and to aid camera operation. For further details see Chapter 22.

When high-sensitivity cameras are used in such widely ranging lighting conditions, from bright sunlight to artificial light, automatic control of light intensity reaching the camera target is needed to avoid blinding or overloading it.

Automatic camera light control

This is achieved by reducing all incoming light to the lowest level the camera can work with satisfactorily, and this, in turn, is done by using an automatic iris which 'stops down' the lens in just the same way as on a film camera. Yet further attenuation is achieved in automatic gain control of the camera's electronic amplifiers.

Charge-coupled targets lend themselves more readily to all-electronic control of light level than the use of mechanical and optical control.

Indoor and outdoor use

If one can cope technically and operationally with outdoor conditions, there should be little difficulty in designing a system for indoor use. A factor common to both, and common to film and CCTV cameras, is the relatively narrow angle of view of the majority of lenses compared with the human eye. To compensate for the narrow angle of view given on the security officer's monitor screen, it is common practice to mount the camera on a 'pan-and-tilt' head, with a joystick type of control for the security officer to operate to keep an intruder in view. One has to ask though, is it right to try to use security officers as camera technicians using pan, tilt, focus and perhaps zoom controls to search for and hold a situation on the screen? My inclination is to say that it is not right. Security officers have other things to do, some of them in split seconds, and often at the same time as they are being asked to operate camera controls.

Using fixed-line-of-sight cameras

The alternative is to get the security system designer, who may be you, to do the work, before the emergency. Apart from the operational problems, pan and tilt systems are expensive and at least two fixed-line cameras can be provided for the same price.

Two cameras on fixed lines of sight, spaced apart and each looking at the other's dead zone and beneath and around it, coupled to an intruder detection system give in my view the security officer the best opportunity of seeing what is going on, without technical distractions.

There is nothing like technical uncertainty to make a man hesitate, and this would be just at the moment when his confidence is needed to enable him to act positively on what he sees. So, I would suggest that CCTV systems should be built on the fixed-line-of-sight principle, with the widest angle lenses practicable, with low light level camera targets and with lighting switched by intruder detection devices. Pan, tilt and zoom have their place, but as exceptions rather than the rule.

Film surveillance and video recording

With the vast improvements made with the recording of TV picture information, film has to be regarded as obsolescent for security purposes. Recorded TV information needs no processing and can be replayed immediately after recording. Video recording thus becomes a natural part of a CCTV system, particularly when cameras are used intermittently, as when switched by intruder sensors. Video recording is discussed in Chapter 22. Before purchasing video recording equipment, however, it is wise to check the current legislation regarding its use.

Security staff safety and costs

It is worth emphasising one or two of the points raised in the opening paragraphs of this chapter. You need them for your own understanding of surveillance, and you may need them if you have to sell the idea of CCTV to someone.

The focal point of a security system has to be the control room security officer, whether he is on or off the site. High responsibility is vested in him to take the right action at the right time, even if he has never had to take that action before. To do this he needs understanding of what he has to do, he needs confidence in the information upon which he acts, and he needs discipline with supervision to cover lapses of human nature. To give him discipline and supervision may be someone else's job, but to give him confidence is our job.

It is hard to imagine anything better calculated to give him confidence than seeing for himself so that he knows what is going on. Rather than spend annually recurring revenue on sufficient patrol officers to feed him with information, it is better to spend once-and-for-all capital on a good CCTV system to give him first-hand information. This also saves exposing security men to attack when on patrol.

Discussion points

This almost certainly is a controversial chapter, and I can safely leave you to pick on several views expressed as discussion points. If possible, try to set up a discussion between people new to the job and people who have experience of working with and without CCTV systems.

8 Alarm communication and control

Security forces are concerned with preserving the status quo. They cannot act effectively unless they are informed that other forces are at work to change the status quo. Thus, communication plays a dominant role in security. An electronic intruder detector is of little use to anyone unless means are provided for passing information to someone who can take counteraction.

This chapter is concerned with various means of passing and presenting that information, and with methods of control of communication that can influence the risk of false calls being made.

The information to be conveyed

If you took a telephone call which simply said, 'Come quickly', what could you do? If you recognised the voice, you would perhaps be able to guess where you were to go to, and indeed to judge whether the caller was trustworthy, as distinct from being, say, a practical joker.

When deciding what information is to be conveyed in alarm communication, put yourself in the position of those who have to take action on the information. The illustration above emphasises the need for eliminating or minimising uncertainty. To be effective, action does have to be quick, and this is encouraged by confidence that the information is complete and correct.

First, there is a need to know the sort of risk involved, requiring reaction by police, fire or other forces. This distinction is sometimes established inherently in that nothing but intruder alarms, requiring police action, come up on that particular panel, but clear distinction is increasingly important as integrated control panels become more widely used.

The second requirement is information of the location of the incident geographically so that the reaction forces can find it readily. It is more of a 'nice-to-have' than essential to add information on whereabouts in the risk area the incident has occurred. This information can more readily be picked up on arrival at the site.

Regarding identity, this is established almost inherently when the call is from an automatic signalling system using a recorded message, and with such a system valuable time can be wasted in trying to verify the source of the call. Verification of identity becomes essential, however, in communications ostensibly between a site security officer and, say, a centralised alarm station. The message could be 'We are working overtime tonight and we will be closing an hour later than usual'. Taken at its face value without verification, the central station might be unaware if this message turned out to be a cover for a raid to be made shortly after closing time.

Establishing the identity of the caller would reduce the risk. Dealing with the risk of collusion and detailed methods of verification are not matters for this book, and those responsible for security personnel training may regard them as matters for verbal rather than written instruction.

Thus, in alarm communication the information required is very simple and the action required is equally specific – 'Come quickly'.

Communication methods

A moment's thought will illustrate that it would be impracticable to link every intruder detection sensor directly with the reaction forces. Each such device would have to carry its identity and location in any information passed, and there just are not enough separate channels of communication available to do it. Techniques such as fibre optics offer scope for evolution, but it is unlikely that future systems will differ widely from the present general arrangements.

In these arrangements communication for the more serious risks is channelled through two or three of the following three stages:

(1) Sensor to local control.
(2) Local control to central control.
(3) Central control to reaction forces.

Local control is located on the protected premises. If the premises are of the lock-up-and-leave type, the local control equipment would be unattended and would pass on the information automatically to a centralised control station serving many such premises. For some risks the reaction forces, the civil police, may provide the central station service, and the functions of central station and reaction force control are combined.

Similarly, if the premises are attended at all hours by local security officers, the functions of local control and central control might be combined, making local security officers responsible for communication in a prescribed form with the reaction forces.

For lower levels of risk communication costs can be reduced, but usually at the cost of increased time between the alert call and attendance by the reaction forces. Typical lower-cost communication methods includes alarm bells on the protected premises and automatic dialling of selected telephone numbers.

Up to the end of the 1970s the vast majority of alarm installations were interconnected with copper wiring – hard wiring. The wiring of the individual alarm installations would normally be carried out by the alarm company, while communication from the protected premises to the reaction forces' control point would normally be via public service telephone lines, rented for a few minutes when needed, as with automatic dialling machines, or rented continuously as private lines or direct lines for the more serious risks.

From the 1980's onwards, increasing use has been made of radio links on site between sensing devices and local control panels. The trend was accelerated in the UK in particular, by legislation breaking the then Post Office monopoly over communication methods.

Selection of communication methods

It is unlikely that the systems man will be able to decide on the communication methods to be used in any particular situation solely on his own view of technical and operational needs. The police will want to get the villain, and will try to insist upon the local part of the system being silent, to avoid the risk of scaring the villain away before the police can get there. Insurance companies, on the other hand, want to minimise their risk of loss, and want local audible alarms to scare off the villain as soon as possible.

Compromise decisions on such conflicts of interest have to be reached at Trade Association level, in which all interested parties other than the villain are represented. Any decisions reached can then be incorporated into alarm company policy and adopted as appropriate into system proposals. We are all concerned with crime prevention, and whether one supports the police or the insurance viewpoint depends on whether one's interests require prevention of the crime to be 'now' or 'next time'.

While on the subject of local audible alarms, whether they be part of a 'bell only' alarm system or part of a comprehensive central station alarm system, they have fallen badly into disrepute with the general public, owing to incessant ringing whether from false or real alarm conditions. Legislation to impose strict limitation of ringing time for audible alarms has to be welcomed on behalf of the public and of the alarm companies, whose reputation would otherwise continue to suffer.

One alternative or additional deterrent suitable for night-time use is a bright light. No one up to mischief feels too comfortable working with a spotlight on him.

Turning now to distant as distinct from local communication, the police, again, influence events, in that they may require an intermediary between their reaction forces and the alarm system. This has given impetus to the growth of proprietary centralised control stations where proper co-operation with the customer can do much to minimise false calls to the police, particularly at opening and closing times. These aspects are enlarged upon somewhat in Chapter 24.

A point with which the systems man will be concerned in the selection of the communication method, particularly on site, is the relative costs involved. It is painfully clear from an examination of the detailed costing of quotations for installation of new security systems that the cost of labour on site for installation is an increasing proportion of the total cost of the system. Custom-installed hard wiring on site runs the risk of pricing itself out of the market as the advantages of using radio become appreciated by system designers.

Don't, however, overlook the dual question, the second part of which is, 'What can the villain do to make it go wrong?' If he succeeds in jamming radio signals in a way that does not reveal the location of the jammed system, it may be necessary to go back to hard wiring to ensure satisfactory communication. If that does happen, it may also be necessary to continue to put up with what appear at first sight to be petty restrictions. But restrictions on what can be connected to public service telephone lines, and the requirements regarding maintenance, stem from the underlying requirement for safety of others. Danger can arise, for instance, from negligent or accidental connection of mains power supply voltages to telephone lines.

Local control panels

Why not leave it switched on all the time? Oddly enough, most intruder detection systems are left switched on all the time, and the only bit that might be switched off is the communication link between the detection system and the reaction forces.

It is obvious that in most industrial and domestic alarm systems the intruder detection facility is needed only when the building is unoccupied. Once people are in the building, legitimately, the detection devices would be triggered so frequently that without the means to switch off the communication link the reaction forces would be receiving an intolerable stream of false alerts. But there is one notable exception. Perhaps because of the success of intruder detection installations during the silent hours, villains are finding it more profitable to attack when people are around, people who can be intimidated into giving the villains what they want. To counter this, the 'panic button' or personal attack switch is provided increasingly in domestic and industrial systems to call the reaction forces as unobtrusively as possible during daytime attack. The personal attack switch by-passes the main control switch that isolates the detection sensors from the communication link to the reaction forces during normal occupation of the building.

Apart from that exception, local control needs in essence only one switch. Why, then, the sometimes quite comprehensive-looking local control panels? The reasons are several, some reasonable and one or two questionable.

The most reasonable justification for multiple switching is found in multipurpose buildings, where one or more parts of the building remain in use by, for instance, evening or night shifts, while other parts of the building are closed for the night and need the intruder detection system to be switched on.

From the alarm company's point of view, multiple switching, or zoning, is justified on the grounds of ease of maintenance and locating and isolating faults should they occur. The local control panel is one of the few places where the alarm company can focus the mind of the customer on their skill in layout, and ease of using the controls. If the supplier buys in all the remainder of the equipment needed for an alarm installation, most alarm companies make a point of supplying control panels to their own design and, if possible, of their own manufacture.

However, it is not unknown for some alarm companies to supply rather more comprehensive control panels than are strictly necessary, partly, perhaps, to increase the sale value of the installation and partly to impress the customer.

Indicator panels

A very valid reason for zoning, particularly in a large installation, is to indicate to the reaction forces the location of the intrusion so that minimum time is lost after arrival in reaching the area at risk.

Please note the operative phrase in the last sentence – 'to indicate'. You do not require switches to provide zone location indication. Indicator lights can be associated on the control panel with each zone switch, or they may be mounted separately on a plain panel, or on a mimic or graphic panel showing the layout of the building or site. The value of the mimic panel lies in its ability to convey accurate information to the control room operator or to the reaction forces in a form that can be easily understood and acted upon.

It will be evident that the same indicators can be used also for fault location; the need for switching then begins to look rather thin. In Chapter 24 it is suggested that parts of a building used in different work patterns are inadequately protected unless each work zone is treated as a complete security risk on its own, with its own alarm control facility and final exit door.

Given this understanding, and before following the indicator panel line of thinking any further, we need to distinguish between a fault arising from, say, a door inadvertently left open which can be cleared by closing it, and a system fault arising from accidental or malicious damage to cables or equipment, or to equipment malfunction, which will need attention by a service engineer to clear.

With this distinction in mind, it should be easier to separate out in control panel design the functions of indication, operational zoning and fault isolation. It should also be easier to avoid adding unnecessary features to the control panel which lead only to operational confusion, to the risk of false alarms and to impaired equipment reliability.

Overattention to the control panel is also a liability, in that it attracts attention from the villain. Hopefully, gone are the days when a control panel could be put out of action by wedging open the moving parts of relays with matchsticks, but scope for mischief remains.

False alarm risks

We said at the outset that in essence all that is needed is a switch to link the detection system to the communication system. Look at a typical control panel and, among all the other switches, you'll probably find a switch that does just that. Unfortunately, too many control panels rely upon operator discipline to prevent a false alarm being created by switching on while a fault exists on the system, and although many expedients have been devised to minimise the problem, only some form of interlocking seems capable of dealing effectively with human fallibility.

Even with a perfectly designed control panel, however, it is salutary to realise that the job is not yet done. The operator still has to get out of the building and lock up behind him.

That creates a further problem. Either that door, usually called the final exit door, has to be excluded from the intruder detection system or it has to be included in such a way that the operator has time to get out before the . . . yes, you've seen the dilemma before I can use all the words needed to express it.

A typical expedient is to use an automatic time delay between the detection and communication systems, set for, say, ten seconds, time enough for the operator to set the control panel, leave and lock up the final door of exit. But what happens if on his way to the door he suddenly realises he has left his keys or whatever in his office? Back he goes, forgetting all about the control panel. The time delay expires while he is getting his keys; the system decides that he is an intruder, and there goes another false alarm. No wonder the police insist on something better.

The final exit door

That better way can only be achieved by making it impossible for the detection system to be connected through to the reaction force communication link until the operator or keyholder, the last to leave the building, has indeed left the building and locked the final exit door. A compromise might be for the local bell alarm to be switched on, with a time delay, at the control panel as before, while retaining the essential requirement that the police calling circuit can only be put through after the operator or keyholder has left the building, and, of course, the circuit must be cut if he goes back in again for whatever reason.

A way that I have used whenever practicable is to install the police calling circuit switch in the keep of the final exit door lock. Those who are familiar with the conventional shunt microswitch in the mortice deadlock of a final door of exit should note particularly that I have said 'in the keep'. There is no way I could encourage the conventional method, and the reason again is the real risk of false alarms at closing time. The keyholder has only to operate the lock with the door open, perhaps to check that he is using the correct key, for a false alarm to be triggered. Further details are given in Chapter 24.

With the keep method, properly engineered, there is virtually no risk of a false alarm at locking-up time. A further merit is that in nearly all but the largest systems the operator ceases to need a control panel, and where it is needed for engineering reasons, it can be installed in a protected area, perhaps remote from the final exit door.

It has been argued that putting such a crucial switch in the final exit door frame is too much like putting all one's eggs in one basket. If doing so increases the prospect of the police giving prompt attention to a real intrusion, instead of dealing with someone else because they think yours is just another false alarm, the risk surely is worth taking.

And the risk can be reduced further by using a final exit door, a new one if necessary, properly designed to resist physical attack and equipped to raise an alarm itself if attacked so vigorously that a break-in or damage to the lock switch could occur. Door design is discussed further in Chapter 23.

Others will argue that it has all been done before. So it has, but why, then, are there so many troublesome control panels and door exit systems still in use, and why are there still so many false alarms at closing time?

In their efforts to reduce false alarms, the UK police issued a document in 1995 called 'ACPO Intruder Alarm Policy'. This policy made clear on what basis the police could refuse to attend alarm calls. The principal effect was to withdraw police attendance at any premises where the number of false alarms exceeded a specified rate. As intended, this had a dramatic effect upon alarm system providers, and much of their work was concentrated on the final exit door – an area already established by the false alarm figures as the main source of trouble. As a means of really understanding the false alarm risk situation, it could be instructive to read again the paragraphs above following the heading 'The final exit door', written in 1988, and to compare them with the measured being taken by system providers to avoid falling foul of the ACPO sanctions.

In the risk business there is no one answer to all problems, and decisions are influenced by the relative importance given to different aspects of the problem. Also, what is good practice today may be obsolete tomorrow, or in ten years' time. What matters is that we understand the principles as thoroughly as we can, and then we are equipped to make sound practical decisions for each situation as it arises.

Discussion points

In this chapter we have treated communication and control as the vital link between the detection system and the ultimate users – the reaction forces. There is so much scope for discussion of ways and means, pros and cons, insurance, police, false alarm control, user and neighbour attitudes that the reader might like to make his own list of discussion points to raise at any and every opportunity with colleagues and others until he feels really familiar with a many-sided subject.

9 Reliability: cause and control of false alarms

Detection is easy – the skill in security work lies in avoiding false alarms. How do we acquire that skill? The fact that false alarms are still such a problem suggests either that no one really knows how to deal with them or that people do not care, enough, to solve the problem.

As we have found elsewhere in this book, it is even more difficult to solve a problem unless you can understand just what the problem is. So, in this chapter we shall try to gather together points made as they cropped up in other chapters, and add other points in an attempt to help you to define for yourself what you believe the false alarm problem to be. From these, together with other like-minded people, you should be able to draw on your understanding to help to get rid of the scourge of false alarms.

To make a start: what is a false alarm? From the intruder's point of view it is any alarm which is not caused by an intruder. Put another way, it is an alarm which results in reaction forces attending premises unnecessarily.

Some designers and alarm company representatives may argue that this is too broad a definition, and they may try to introduce sub-divisions. They may, for instance, say that the alarm was caused by a mistake by the user and therefore the alarm was not false. A remark on these lines is partly defensive, and from an overall system point of view it is really part of the answer. Should not a designer make it impracticable for that kind of mistake by the user to cause an alarm? If we believe in the objective of preventing false alarms, we have to resist restricting the definition, and we have to say that the designer should deal with the human problems.

By accepting a broad definition we avoid attaching blame, and can focus attention on another definition we need: what do we mean by the system in this context? This definition, too, has to be comprehensive, for if anything is left out, a whole family of false alarm causes may be overlooked. In this context the system consists of:

The environment.
The users.
Interconnections.
The equipment.
The reaction forces.

All are sources of false alarms (including the last) and systematic attention to each will contribute to an understanding of the problem as a whole and to solutions which if implemented would largely eliminate the false alarm problem. As we have seen, the consequences of not dealing with false alarms include possible neglect by operating personnel of a real alarm and withdrawal of reaction force services, with the consequential loss of insurance cover.

How can we know when false alarms have been reduced to an acceptable level? If we are to make a determined attack on them, we need some way of measuring success or failure, and some way of comparing our efforts with the efforts made by others.

A questionable method is the way in which some reaction forces express the overall performance of intruder alarm systems. They say, for instance, that of all alarm calls nearly 90% are false. This is the same thing as saying that, out of every 100 alarm calls, little over 10 are from actual intrusions. Now, suppose that because of an intense crime prevention publicity campaign on the value of intruder alarms, the number of attacks by villains is halved and the number of false alarms stays about the same; then the false alarm rate would be given near enough to 99%, a worse figure for a better situation. At least, to the public at large the figure is worse, but the alarm industry would know that they had made no contribution, neither for worse nor for better.

A ground on which the alarm industry can object with more justification to the percentage method of expressing false alarms is that the method is insensitive. Suppose that in a given time there are 2 intrusions and 98 false alarms; then the false alarm rate is given as 98%. Now suppose that after great efforts the number of false alarms are halved, to 49, and the intrusions remain at 2. The false alarm rate is $49/(49 + 2)$ or 96%. Only 2% reduction for all that effort? It is discouraging, but it does emphasise the drastic reductions in false alarms that have to be achieved to make any significant impression on the percentage figures. And that is the truth of the matter.

Sadly, no simple statistic can include a measure of crime prevention effectiveness. It is likely, therefore, that for publicity purposes, and as a stick to be wielded by the reaction forces at security companies, we are stuck with statements such as '90% of all alarm calls are false'.

For operational purposes there is a better way. Many reaction forces take real intrusions out of the equation and concentrate on the false alarms themselves. Instead of expressing the rate of false alarms as illustrated above, they compare the number of false alarms with time, and this gives the false alarm rate per year. Thus, any individual installation can be given a target rating of, say, two false alarms a year, and if that rate is exceeded, reaction force cover will be withdrawn from the premises concerned. It is tough, and it can be arbitrary, but it is a necessary approach.

This, in effect, is the method adopted in the UK Association of Chief Police Officers for control of alarm system providers, in the ACPO Policy document issued in 1995.

Turning again to the overall system, we can study some of the causes of false alarm in each of the categories already listed. The causes given are only part of the picture and serve as a guide to building up your own more detailed list as your experience grows.

The environment

The only thing we want to detect is an intruder. Out of doors intruder detection devices are stretched near to the limit in trying to distinguish between intruders and legitimate movement by people, animals, birds, vegetation and wind-blown inanimate objects. Weather problems include rain, snow, fog, ice, thunder and lightning. It is a measure of the success of development of outdoor detection devices that so many of these hazards can be coped with without undue false alarms. However, a local manned security office with CCTV and other aids remains a highly desirable false alarm filter between an outdoor risk and the reaction forces.

Indoors the environmental problems are much more readily controlled once one is aware of them. Warehouses and the like are borderline cases, where birds and cats are typical hazards. Movement of air from radiators and overhead heaters, movement of fan blades, rattling of doors and glints of sunlight are more generalised problems.

Coastal areas and some industrial processes cause corrosive atmospheres which in time damage equipment sufficiently to cause breakdown, and in a fail-safe system that means a false alarm.

In areas of high temperatures and humidity fungal growth and insect infestation add to environmental factors to be taken into consideration.

The users

The fact that so many false alarms occur when premises are being opened up or locked for the night is an indication of the effect of users upon the problem. Whether it is the result of checking in the lock for the right key, going back into a protected area after the alarm was set, leaving a door or window open, or inadvertently pressing a personal attack button, the false alarm is the same.

Interconnections

Copper is a curious material. When made up into wire for interconnection of electronic components and when made up into cable for interconnection of separate items of electronic equipment, the copper is treated in manufacture to have a certain hardness, not too hard and not too soft, but just right for easy handling. It can be bent to shape as required, and once installed it is as reliable as can be.

But as the copper is bent, it becomes slightly harder. Bend it again later, accidentally, perhaps, when a service engineer is checking something else, and it becomes harder still. A stage is reached in this process when the wire becomes brittle, and further bending can cause it to break. For cabling, in particular, it is better to use conductors made up of several strands of fairly fine-gauge wire than to use a single strand of rather thicker wire.

Other problems in interconnection include staples used for fixing cables to walls, etc., accidentally driven through the cable in such a way that when the cable moves with temperature changes, unwanted contact is made between adjacent conductors.

One of the worst offenders is the connector block where cables are joined to other cables or to equipment. In some types of block the 'pinch-screw' makes copper brittle in just the same way as bending does, with the same disastrous results.

To control the problem when the type of connector is appropriate, only connectors having a 'leaf', usually of phosphor bronze, built in between the screw and the cable should be allowed by security installation and service providers. This can mean rejecting bought-in equipment unless it has suitable connectors.

The equipment

It is not so long since when questions of reliability and false alarms came up, the first thought in one's mind was for the equipment in the system. Problems ranged from contacts in mechanical relays to heat-accelerated damage due to adjacent inefficient electronic components.

Progress in the use of semiconductor technology has been so rapid and effective that electronic switching has taken over from mechanical relays, and overall efficiencies are so much higher that waste heat does little or no damage, yet is just sufficient to keep components and equipment free from damage by moisture. The effect is that the equipment has become the most reliable part of an overall intruder detection system. Failures do occur, however, and there is no room for complacency from anybody from designer to service engineer if equipment is to maintain its record as regards contributing to the false alarm problem.

Reaction forces

It may seem strange to include the reaction forces such as the police in a discussion of the causes of false alarms. They may not in fact be the cause of a false alarm itself, but they can be the cause of an incorrect report of a false alarm against a particular installation, as the following incident will show.

Police and keyholder attended a warehouse following an alarm call. The building was searched very thoroughly, but there was no trace of intrusion and the search was called off. When the police were assembled outside ready to go back and report that it was a false alarm, an alert officer noticed a noise from an unexpected direction. Something prompted him to go on to the roof of the building, and there, in the trough between two skylights, he found a man who later admitted that he intended to remove a pane of glass. What neither the man nor the police realised at the time was that the warehouse was equipped with a device known as a 'radio alarm' from which UHF energy leaked through the glass of the roof and detected the potential intruder. The alarm was real, but it might well have been logged as false.

Other illustrations and remarks about false alarms appear in various parts of the book as the need arises, sometimes in the guise of nuisance alarms, and following these up is a good example of the use of the index. Even then the story is not complete, but when they are added to your own experiences you will have a very fair understanding of the subject and, hopefully, you will have an urge to do more than something to overcome the problem.

As applied at the time, the UHF radio alarm suffered the same limitations as the standing wave version of the ultrasonic detector described in Chapter 15, and although it was good at detection, particularly in large buildings such as warehouses, it was also prone to giving false alarms. Had we known then what we know now, that could have been overcome so easily.

The tendency then as now as to discard the troublesome in favour, perhaps, of some new device or in favour of an older less effective but less troublesome device. Adverse experience in the field is infectious, and leads to fashions and, worse, to prejudice in the selection and use of detection devices. Too often a device gets a bad name because it was used incorrectly and this comes back to training and to care in installation design.

Building confidence and selling quality

To be fair, though, someone has to be persuaded to buy the system, against competition from others. To do the installation properly may cost more in initial costs, even if greater reliability leads later to lower maintenance costs and to fewer call-outs for false alarms. Initial costs are a fact, and lower running costs are but a forecast, and it takes a courageous accountant to believe that paying a higher initial cost will give a lower total cost after five or ten years' running.

An overcautious alarm company will use this customer resistance to 'total cost' concepts to standardise his systems beyond the point of wisdom. Standardisation should be a guide for the time being and not a barrier to evolution. The progressive company will ensure that there is consistent feedback of information regarding the performance of equipment and of systems in customers' installations. Field experience would then be used by the equipment, system and installation designers to improve their products and their work. This approach not only improves their own confidence, but also generates confidence with the customer, the only sure way of selling quality at higher prices. 'Toffee shop' business has its place, but not every firm wants it.

Given the confidence, it is easier to offer maintenance and call-out contracts which reflect the higher reliability and lower false alarm rates which follow from proper use of field experience fed back to the design departments. In this way it becomes practicable to offer, and not just to forecast, lower total costs – an argument much more likely to influence the customer's decision in your favour.

Discussion points

You will remember that the total cost concept covers the equipment costs, installation costs and running costs, which include routine maintenance and call-out service over a period of five or ten years. If reliability is improved and false alarms are reduced, then running costs are reduced. Try to set up informal or formal discussions with accountants and others with financial preoccupations to get other points of view and to find areas of agreement that can be used to guide future decisions.

Discuss with colleagues any experiences with reliability and false alarms that can help to focus attention on the problems. Although implementing solutions in this case is harder than in general because of dependence upon various other people, aim to use the accumulated information to inspire 'these-things-we-will-do' decisions.

10 Movement control of personnel and materials

We are such polite people: we will even hold a door open for a stranger following us into a building. To resist this good will and other less well-intentioned motives it has been customary from olden times to assign a door-keeper to look out for and to bar entry to anyone who might be unwelcome within the building.

Access control – the objective

The practice continues to this day, with perhaps the name of the function changed to usher, commissionaire or security officer. For the doorman to discriminate between friend and foe, the function implies knowledge and training, and it also implies having sufficient physical means for preventing entry when required.

When a firm examines its payroll, it is not easy to quantify the value of the function of door-keeper, and it is not surprising that from time to time the question is asked: Can the job be done in another way? Surely electronics could save labour costs here as they have in so many other functions?

An incomplete answer

Again, not surprisingly, electronics people were only too willing to oblige, and the market was flooded with a range of electronic access control equipment which went a long way towards satisfying the requirement for selecting only acceptable people for access to buildings and parts of buildings. By its nature, however, this work attracted the newer digital breed of electronic designers, who, with respect, had not then learned, as their analogue brethren had learned, the security dual question: 'Will it work? Can it be made not to work?' In other words, they provided sophisticated electronic control of the door-lock, but inadequate attention was given to the physical means of preventing access by a hostile intruder. Without solutions to that problem there was no way that access control could replace satisfactorily and save the cost of the door-keeper.

Adding door control

The incomplete nature of the electronic answer arose somewhat on these lines. When you buy a key to control a door, you can hardly avoid buying a matching lock with it. So, at one act, you buy both the mechanical part for holding the door closed – that is, the lock, which really is a key-controlled bolt – and the means to control it – the key. On the other hand, if you go out to buy access control equipment, what you get in effect is only the key, and quite separate decisions and purchasing actions can arise when it comes to getting the lock or door-bolting mechanism.

Once this is understood a wider range of choice is available. For purchasers this can be a good thing but for access control equipment designers it is a bad thing. They tend to concentrate on the electronics and remote control aspects of the job, and to regard the mechanics of door design and actual locking and bolting of the door as the job of someone else.

This split of responsibility can lead to inadequate attention to ways of dealing with evasion – that is, with successful unauthorised entry. Can some access control systems, then, be charged with being technically sound but operationally useless?

The charge would be invalid provided that overall system engineering concepts were applied to co-ordinate the use of the various components that go to make up an operational access control system. What has happened, owing to the wide range of skills involved, is that a separate discipline of door control has had to be recognised. Door control is concerned essentially with the mechanical bolting and locking aspects of implementing access control, and consequently draws within its ambit the selection of door type and construction, including its frame and methods of securing it to the surrounding walls.

A dilemma remains

When I first had to face up to the realities of the automatic control of access, I believed I had mastered the situation when I had thought through the need for the door control concept in addition to access control. However, a very uneasy feeling remained that automatic control was not being achieved, despite this being the object of the exercise. When I feel confused in this way and uncertain of the answer (or, more truthfully perhaps, knowing the answer and being afraid to admit it because of the consequences), I sometimes write down a step-by-step analysis with the aim of confronting myself with a situation I have to accept. I did just that for access control, and as the method and results have some value in training, I am repeating the analysis below, deleting only some of the confusions and contradictions encountered on the way.

Analysing the problem

I have called the early stages of access control a cult, and have drawn upon myself a fair amount of flack for doing so. Others have called the problem a dilemma, but the best word was used by a colleague who understood it better than anyone I have discussed it with – he called the access control situation a paradox. If you are wondering how the paradox arises, I should like to invite you to follow through a series of comments, questions and answers aimed at isolating problems and at putting together solutions.

The origins of access control

The objective is to allow access automatically to authorised personnel and to exclude automatically non-authorised personnel. Behind this objective lies the requirement to eliminate the need for manpower for the control of access.

What is meant by access?

It means the movement of personnel from the outside to the inside of a building, or from one area within a building to another area within the same building. The concept of access can include also the movement of people in the opposite direction, from inside to outside. Access control does not mean the control of unaccompanied materials, nor does it mean, although it can include, the control of intruders during the silent hours, when alternative or additional restrictive measures may be taken.

What is meant by control?

A prerequisite to control is a decision on the question 'Which personnel are acceptable within the risk area?' The key decision that follows from this is that all other personnel are to be excluded from the risk area. Given these decisions, control is that by which decisions are implemented. Implementation is by discipline, by physical barriers and by electromechanical aids.

Can control be absolute?

Security is essentially a risk business. Complete success would in all probability cost more than can be spent. Anything less than complete success means accepting the risk of failure. Acceptance of the risk of failure means that a choice has to be made on what to omit. That choice is amenable to analysis, but sooner or later human opinion has to be exercised in deciding what combination of weighting factors applied during analysis can be accepted. Different individuals and different groups of individuals may arrive at different decisions on what measures to take and to leave out.

What is wrong with a lock?

What is wrong with a door? Is there a need for anything better? Finding accurate definitions for the words 'lock' and 'door' can divert attention, so the colloquial meanings are assumed for the moment. Further questions are necessary before a credible answer to this question can emerge.

Can the lock be released from outside?

If no provision is made for opening the door from outside, then, subject to the quality of the door and the lock, the door while locked closed becomes in effect a continuation of the perimeter wall and gives a similar degree of security against unauthorised entry.

If no provision is made for opening the door from outside, how is access to be achieved? If by bell, does someone go to the door to identify the person outside via, say, a peep-hole? For the conventional lock and door situation the answer is likely to be yes, and has the merit that positive visual identification of the person is possible without first opening the door.

Up to this point, there is little or nothing wrong with the conventional lock and door arrangement, but beyond this point a further set of questions will illustrate the changed attitude to conventional methods.

(1) Who does go to the door in answer to the bell – anyone in the department, allocated personnel in the department or security staff?
(2) How long does it take to allow entry or to refuse entry?
(3) How much is it likely to cost per annum to provide this door attendance?
(4) If these questions raise doubts regarding the continued viability of having even part-time door attendance, what are the implications of other arrangements?
(5) For instance, what are the implications of providing external key release for the door-lock?
(6) How many 'authorised user' keys are required? How many doors are to be controlled? Are all doors to have the same key or different keys?
(7) Can a key be obtained by an unauthorised person, through loss, duplication or collusion?
(8) Can an unauthorised person with a key gain entry unobserved?
(9) If door attendance is rendered unnecessary by the adoption of keys for external access, who closes the door after authorised entry?
(10) Who would act if the door were found to have been left open?
(11) If an automatic door-closer is fitted, can an intruder follow an authorised person through the door before it closes?
(12) Can the automatic closer fail, or be cause to fail, to close the door completely?

More questions can be posed, but the pattern is set for anyone to add their own if need be. For many, however, it will be clear that there is no simple alternative to a door attendant or, worse, to allowing free access to anybody. But in modern industry the cost of manning, the cost of delay caused to authorised personnel, the irritation caused through not being able to continue on one's way and the consequences of allowing unauthorised entry all lead to the call for something better – in fact, for access control.

So far the analysis has done little more than restate the problem rather than solve it, so we must try harder and see what happens when we do have access control.

Some basic requirements

Just recapitulating for a moment, the concept of access control requires that it shall not be necessary for an attendant to identify and admit an authorised person. Thus:

(1) Identification must be automatic, and there is a reasonably wide range of suitable electronic devices available for achieving this, as described in Chapter 25.
(2) The identity of each individual must be established before admission. This means that the door must close and be locked after each single admission. There is no way in which this requirement can be relaxed if effective automatic access control is to be achieved; hence the complete dependence of electronic and electromechanical controls upon the physical design of the door as a barrier.

Physical designs of door

A list of possible designs for personnel access might include:
Single-leaf side-hinged swing door with automatic closer.
'Ticket barrier' turnstile.
'Hotel entrance' revolving door with indexed lock for each 90° of rotation.
'Zoo entrance' interleaved steel bar revolving door, indexed as above.
Two single-leaf doors in sequence, swing or sliding, spaced the minimum practicable distance apart to form an 'air-lock' entrance in which the doors are interlocked to ensure that one door is locked before the other can be opened.

Features of various types of door

The target objective again is that the door must close and be locked automatically after each single admission. Using a check list, the features of different door types become clear.

(1) For swing doors:
Can the door be left or wedged open? If yes, does it matter? If it does matter, security staff or others need to be alerted automatically that the door is open, and an attendant is needed to clear any obstruction and to close and lock the door.
(2) For ticket barrier turnstiles:
How easy is it for an intruder to evade it? If evasion cannot be prevented, then security staff must be alerted automatically so that they can deal with an intruder.

(3) Hotel entrance and zoo entrance revolving doors have the merit that they cannot be left open, and if rotation is maliciously jammed, security is not weakened and the next authorised user may remove the obstruction himself rather than relying upon an attendant. But are revolving doors too space-consuming and expensive?

(4) For air-lock entrances:
Would entry through two single interlocked doors be sufficiently rapid at personnel arrival and departure times? If yes, they have the merit of revolving doors that the doorway cannot be left open, and the need for attendance is minimal. If no, the relatively low cost would not prohibit the use of more than one air-lock entrance door.

(5) Visitors:
But even now we still have the problem of maintenance people, outside contractors and business visitors. They need to be able to announce their presence at the door, and an attendant has to decide whether or not to let each person in, or to reject them, and to act accordingly.

Accepting the paradox

Here, then, is the paradox. Access control is intended to eliminate the door attendant, yet again and again we are finding that, even with access control in operation, the attendant is still needed.

If access control is intended to be an effective safeguard against unauthorised entry, then all concerned need to understand that access control can only reduce, and cannot eliminate, the need for door attendance manpower. As described in Chapter 25, if access control equipment concentrates on dealing automatically with acceptable people, and on referring the exceptions to the security staff, it is possible that sufficient savings on manning can be made to more than pay for access control and associated door control equipment.

The paradox is real, but acceptable. So, what access control systems are there?

Access control systems

Who will want to get in? Authorised personnel. Acceptable but unauthorised personnel – visitors. Unauthorised personnel – villains.

Proving authorisation

How can a person prove to the system that he is authorised?
With a push-button mechanical lock operated from a memorised code.
From push-button operation of an electromechanical keyboard using a memorised code.
With a mechanical key operating a mechanical lock or operating a switch linked to an electromechanical lock.

With a plastic card carrying a concealed identity code for insertion into a card 'reader'.
With a code-modulated radio, ultrasonic or infrared short-range transmission.
With a plastic card coded to influence a nearby sensor responsive to magnetic fields or electrostatic charges.
With a combination of key or card systems and push-button systems.

Choice of access control system

All the above methods of proving identity can and do work, and the choice of available methods is constantly increasing. But the dual question has to be asked yet again: 'Will it work? Can it be caused not to work?'

Unfortunately, there is much that a villain or a vandal can do. For example:

(1) *Push buttons*. If all authorised users of a mechanical push-button lock use the same memorised code, the risk of an unauthorised person learning it by accident or by skill is high. The risk increases with the number of authorised users. Frequent if irregular changes of code are self-defeating, as too many users fail to remember the current code and an attendant is needed to rescue them.

(2) *Mechanical keys* can be lost, stolen or duplicated. The risk is minimised if authorised keyholders are obliged to surrender the keys each time they leave the premises. But can this overload the working of the security office?

(3) *Plastic cards*, like keys, can be lost or stolen, but are more difficult to duplicate.

(4) *Keyboard-controlled electronics* enables each authorised person to have his own identity code, instead of using the door's own code, as in the case of the push-button mechanical lock.

(5) *Combined plastic card and keyboard control*. The cross-checking feasible between identities established by both plastic cards and keyboard codes raises the security against fraudulent entry, but mechanical weaknesses remain.

(6) *Card readers* frequently have slots into which the plastic card is inserted for identification. The slots, like keyholes, can, unfortunately, be plugged with chewing-gum or the equivalent to prevent or delay proper use, and attendants are needed to clear them and to monitor access meanwhile.

Some card-readers defy villains by avoiding the use of slots. The card is presented face-down to the reader device instead of edge-on, and there is much reduced risk of obstructions preventing the card being read.

(7) *Proximity identification*. A near approach to vandal resistance is given by a door with no keyholes, no slots and no external means of opening it. An authorised person can release the door by means of a proximity device using radio or inductive coupling to an internal mechanism. As it provides 'hands off' operation, authorised access is quick, and attractive

for situations where traffic density is high at certain times of the day and for randomly timed movements at any required part of the day.

Some organisations dislike having the pocket proximity device taken home at night on security grounds, and indeed some devices have to remain on site overnight for battery charging. No attention by security personnel is needed while authorised personnel deposit their pocket device in a pre-allocated receptacle for overnight charging, but security would be alerted automatically if anyone attempted accidentally or otherwise to take a device off the site.

(8) If a viable device for verifying physical characteristics of face, hand or voice, say, can be devised, perhaps we are getting somewhere in frustrating the villain – Chapter 25 points the way.

(9) *Implementation.* We set out to analyse the access control problem, and in the process we exposed the paradox that it does not after all eliminate altogether the need for a door attendant of some kind. We also isolated the separate function of door control, without which access control is ineffective. Most important, we accepted that anything less than perfection means taking risks.

Good door and access control can start with a single swing door with an electrical lock remotely controlled by a part-time attendant who can see the door and anyone wishing to enter.

The analysis helps to highlight the alternatives available and the pros and cons of each. It is a matter for judgment how much risk can be taken and how much can and should be spent on the system to achieve a net saving on manpower or a net increase in security.

This chapter should provide the background against which these decisions can be made, and Chapter 25 provides practical guidance on implementation.

Movement control of materials

Practically the whole of this book is devoted in one way or another to the access control of intruders. Intruders usually bring some material in with them or endeavour to take something out with them. Outside working hours it is normally sufficient to detect the intruder himself, but during working hours there can be so many people around legitimately that it is even more difficult to isolate the person who has evaded the access control system successfully, or who is nominally 'authorised' but is intent on moving materials to his own rather than to the owner's advantage.

For bulky material the same precautions taken with revolving doors and air-lock doors to prevent more than one person being accepted on a single identification should make it that much more difficult to get the material through. For non-bulky or small quantities of material, access control and door control as such are of little or no help, but when combined with physical search and the electronic detection devices described for the purpose in Chapter 26, the risk of loss is reduced but not eliminated, and there is no alternative to the use of security manpower to help close the gap.

Discussion points

Access control is not the panacea some hoped it would be. But despite the problems, it is worth devoting the time needed by a process of elimination to decide upon the system most appropriate to any given situation. To help crystallise the factors involved in your mind, select premises and areas you know, find out the legitimate movement patterns, and discuss with colleagues your ideas of good access control systems for them.

11 Espionage and countermeasures

If 'all knowledge is useful' is the motto of the industrial spy, 'be aware' has to be the motto for the victim. The purpose of this chapter is to alert the reader to the reality of espionage in his own personal and business life, not just in the life of some distant Government agent or in the life of some fictitious 007-type character. Once he has been alerted to the situation, it should become a lifetime habit of a security man to be aware, constantly, of the risks of espionage.

Electronics has its place, as we shall see, in espionage, but in many cases espionage is so simple that electronic techniques are just not necessary.

To focus our minds on the essential aspects, espionage we will say is the gathering and transfer, secretly, of information from one person or organisation to another. It is all concerned with information and its transfer without the knowledge of the owner of the information.

This form of theft is not a criminal offence in some, particularly English-speaking, countries, and consequently it is that much harder for security people to deal with. As for motive, human nature is adept at exploiting curiosity to promote its objectives. From the earliest times there has been the need to eat. This generated in man animal techniques for stalking edible prey, and camouflage techniques in the prey to conceal information regarding their location. In more 'civilised' times we employ the butcher to do this work for us, and we have time to take over another company, to add to our prosperity, or to protect our country by finding out details of a new weapon being developed by a potential enemy country, or perhaps we have time just to find out what the firm is paying an office colleague.

If we can accept that the motive is there, in whatever form it takes, let us look at some of the forms the information can take. As I am writing this, it is perhaps natural to consider written information first.

Written information

When published, what I write is my copyright and has a value to me. If some villain thought he could make a fortune by stealing and publishing my words of wisdom under his own name, what scope has he, potentially or in reality, for getting the information he wants?

Written information invariably has a top copy – the original – and may well have a carbon copy. If the carbon paper happens to be reasonably new, the information is available on the carbon paper itself. Also, as writing on a hard surface is rather uncomfortable, one tends to write on a blotter or, as in my case, on a pad of several sheets of writing paper. So, at least one of the sheets in that pad also bears the impression of the writing, and with care and patience the text can be extracted. If a fair copy is made of a rough draft, the draft probably goes into the wastepaper basket, available to anyone who can get it.

Typed information

Typed information faces the same hazards as written information, together with others that are easily overlooked. Businesses that are quite careful about the security of written and typed material may fail to realise the consequences of using typewriters in which the ribbon traverses only once before requiring replacement. The typist, unless instructed otherwise, is quite entitled to throw the old spool of ribbon into the wastepaper basket, unaware or unconcerned that all she had typed was spelled out clearly on the ribbon like a telegram.

If carbon copies are less fashionable now, that cannot be said of photo-copies – a booming business both for legitimate use and for the villain.

Stored and recorded information

The problem extends yet further, and electronics becomes intimately involved not only with dictation machines and tape recorders, but also with word processors and the ever-growing memories of computers and other electronic data-processing equipment.

Spoken information

The chance remark is perhaps the hardest to deal with. Spies have to be ready for it whenever it is made. But there is evidence enough that they are ready, and wartime warnings about careless talk and that 'walls have ears' were based on bitter experience. In comparison, the board room and conference room are easy to penetrate. Advance information is available regarding meeting times, and spies have ready access to all the electronic equipment they need.

So much publicity has been given to telephone tapping that it is hard to believe that anyone can be unaware of the hazards involved. If they are aware, they console themselves with the belief that at least their telephone has not been tampered with. Even if it had, they say with some justification that it is probably physically impossible to monitor, abstract, assess and act upon all conversations. As has been said so often, security is a risk business, and most people take the risk of interception of their phone calls. Again, awareness is the key word, to be heeded when dealing with sensitive material.

Countermeasures

Perhaps enough has been said to indicate the range of opportunities for espionage in Government, in industry, in commerce and in private life. So, what countermeasures can be taken? This is not the place to answer that question fully, nor is it possible, since methods change and counter-measures change to match. The objective has to be to keep one or more steps ahead of the spy.

But some of the so-called common-sense countermeasures can be touched upon on the basis of self-help. In self-help countermeasures your own guideline can be based on the need to know. Get answers to the question, 'Does it matter if Tom, Dick or Ivan gets to know about it?' Mostly it doesn't matter, and you can concentrate on the relatively few items that do matter.

(1) *Filing* is a typical problem. Some organisations label papers and files as 'secret' or 'confidential', with the rest unmarked. If I wanted to keep this manuscript secret, would I mark it as such and put it in a special file, or would I 'lose' it in with a mass of other papers? If you guess from the way I phrase that question that I would prefer to 'lose' it rather than attract attention to it by marking it 'secret', you may be right, or you may be wrong. The danger lies, I think, in having too rigid a system. If the spy knows that such and such a system is always used for filing, maybe it makes his job easier. Some inconsistency has its virtues.

The least one is likely to do, however, is to leave important information in a locked cabinet. If you believe that opening a filing cabinet without a key is easy, you may check the security of the room containing the cabinet, and add physical protection, intruder detection and access control as warranted by the risks involved.

(2) *Outside help*. Working on these lines through the possible areas of risk, you may well come across a situation for which you need outside help, from one of the organisations specialising in countermeasures. How do you find a good one (for there are some that are not so good)? As in most such situations, talk to someone you can trust – maybe in the security business, or in insurance, legal or senior police circles.

(3) *Managerial discipline* has to be discussed, understood, accepted and implemented as self-discipline and as an example to the personnel. It must be the function of management, aided by their security advisers, to establish for themselves, and to issue to their personnel, guidelines and specific instructions regarding habits, routines and emergency actions involving speaking, telephoning, writing, office equipment, laboratory results, production and sales figures, and so on.

Success does not come overnight, but cultivation of awareness filters through in due course, and for those of us in security who have the opportunity of visiting various organisations it is quickly apparent which are making more progress than others towards making espionage that bit more difficult.

Discussion points

If you are involved with security, think back over the last three or four weeks and try to identify situations you know about that could make successful espionage possible. And what could be the consequences? If you can take the necessary countermeasures yourself, well and good; but if the situation needs help from others, discuss it with them or with your immediate superior, as appropriate. The possibilities are so diverse that discussing case histories and current situations probably makes the most vivid impression on the mind.

12 Sound verification

Almost without exception all the intruder detection systems envisaged earlier terminate in a central or local security office as a light, bell, buzzer, bleeper or clear single tone. Security officers are trained in what action to take when they see or hear any of these. Given such an event in a central station, the security officer would normally pass on the alert to the police, with little or no more information than the address of the site at risk.

No amount of training, self-discipline or imposed discipline, however, can prevent human nature intervening from time to time, leading to a feeling that the job is a bit uninteresting. No one is saying that the job will not be done, it will be, but when those concerned know also that 8 or 9 out of 10 such calls are likely to be false, something more is needed.

That something can be sound verification. In essence, such a system consists of a microphone and an audio amplifier in the risk area and a loudspeaker in the security central station, with a dedicated or shared communication link between the risk area and the central station. If the central station officer can now say to the police, 'I heard some breaking glass, then two men talking near the main storage area', or 'I can hear some youngsters running around in the classroom', the police can act with near certainty that the alarm is real.

Figure. 12.1. A 'Sonitrol' sound system installed in a central station security control room.

Probably the leading exponent of this system is Sonitrol, who have installations at all levels, from, say, Fort Knox in USA to schools in the UK. Microphones can be used on their own, but they make a more potent system when combined with any of the detection systems available. As an example, *Figure 12.1* shows a typical Central System set-up, comprising an overall customer indicator panel, a video display showing all the information the police and the security company will need in handling the incident immediately the alarm call comes in. In addition, a print-out of that data is issued automatically for use by those going to the site of the incident, and a recording is made of the sounds being received from the site. Best of all, more criminals are caught in the act this way, and the false alarm rate from sites fitted with sound verification is way below that for 'conventional' alarm systems.

Discussion points

What next? With passive infrared and dual systems and with dramatic reductions in false alarms possible with improved 'entry–exit' methods, will sound verification turn 'burglar alarms' from a time waster into the boon that the police deserve?

Part 2

Equipment

13 Switching sensors

Of all the electronic and electrical sensors now available for detection of intruders the switch is one of the earliest devised, and the fact that it is still used, in various forms, indicates that it has merits that surpass all others for some applications.

What is a switch?

In its broadest sense in the security context a switch can be said to be a device which, when influenced beyond a certain critical condition, can cause a sudden change in the value of an electrical current flowing through it. To be complete, even that definition needs refinement, but it is cumbersome enough as it is, and it serves only to remind us that, really, we already know what a switch is. What comes as a surprise, perhaps, is the many forms a switch can take in intruder detection, and it helps in understanding to distinguish between these and other methods of detection.

The principal virtue of the switch is that it can be made very definite in its action: that it can be definitely 'closed' (current flowing) or definitely 'open' (no current flowing) and the transition from open to closed can be made so definite that it is unlikely to change from one condition to the other by accident. When a device can be trusted to behave consistently in this way then it should be reliable for detection, and the risk of false alarms due to accidental operation should be low.

Making an experimental switch

Simple though a switch may be in principle, it is far from simple in practice. If we want to choose when a current is to pass, the obvious thing to do is to have two pieces of metal which we put in contact with each other when we want current to pass and which we separate when we want no current to pass. The operative word here is 'contact'.

Suppose that we make an experimental switch by using a needle and a coin. If we connect the needle to a battery and the other side of the battery to, say, a car headlamp bulb and the other side of the bulb to the coin, we know that as soon as the needle touches the coin the bulb will light.

Design problems

At least, the bulb *should* light. The first snag is that the coin may be dirty, and only by pressing the needle hard against it to break through the dirt or corrosion can we make the bulb light. Our experiment has shown the first essential ingredient of a practical mechanical switch – contact pressure, to reduce electrical resistance to the flow of current.

The second snag is contact heating. Even with quite high contact pressure there will still be some contact resistance. We know that when current passes through a resistance heat is generated, and in our experimental switch that heat is concentrated into the minute area of contact between the needle point and the coin. It is easily possible in some switches, even if not in our example, for sufficient heat to be generated actually to weld the metals together, so that the circuit cannot be switched off.

Expedients

Designers of switches have to adopt various expedients to reduce contact heating and the risk of welding. In a way, our experimental switch is not too bad, and it shows a little of what can be done.

First, the coin, with a fairly large mass of metal, provides a 'heat sink'. Particularly if the coin is copper, heat is conducted rapidly away from the point contact area, which reduces the risk of the contact overheating and welding up. Second, if a needle made of hardened steel is used the melting point is much higher than that of the coin, and fusion is less likely.

The third snag shows up when the designer tries to increase the area of contact. His thinking naturally says that if the current through one needle point can cause overheating, let us use a number of needles in parallel to increase the contact area. He soon finds himself in mechanical difficulties in doing this. What is needed is similar contact pressure for each of the needles. Try thinking of it this way. It is easy to see that two needles can be balanced to give equal pressure on any surface. Also, as with a three-legged stool, three needles will stand firmly on any surface. On all but a few surfaces, however, a four-legged chair will wobble, and contact pressure on one leg will be much less than on the others. Only by sitting on the chair – that is, by increasing the overall contact pressure – is there a chance of making all four legs share the load.

Even surfaces that look and feel smooth are really a combination of peaks and troughs, as a microscope will show, and making a switch from two smooth surfaces pressed together is equivalent only to several four-legged chairs with their legs under unequal pressures.

So contact heating to a greater or lesser amount at various contact points cannot be avoided. Another expedient is the choice of contact materials. A common one is silver, which has advantages as a contact material because, unlike rust on steel, even the oxide or tarnish formed on silver is a good conductor of electricity, and is not unduly prone to heating, given moderate contact pressure.

So much, then, for the background to switches and an understanding of some of the things to be aware of when we consider them in security applications. Let us look now at a range of applications and the features of switches and switchlike actions used in them.

Door contacts

Already we are using something of the jargon of the security industry. A switch inserted in the frame of a door was probably the earliest form of security sensor, to give warning that a given door had been opened. Switches used in this way became almost universally known as door contacts.

Mechanical contacts

Essentially, in mechanical switches for doors the contacts are enclosed inside a small box, out of which protrudes a plunger which activates the switch-box when pushed. By recessing the switch-box into the stationary frame of the door the plunger is made to operate as the door reaches the fully closed position.

Switches of this type, although often flimsy, could be made robust and with contacts capable of carrying the relatively high currents frequently used in electrical as distinct from electronic detection systems. It is difficult, however, to conceal mechanical switches effectively, and it became traditional to say that they were too easily made useless with a plug of chewing-gum to hold in the plunger. Although true, this was a little unfair, since not all villains have prior access to doors they are interested in.

Reed switches

Where very low values of current and feasible contact pressure are involved, as often encountered in electronic systems, a further expedient to get adequate contact performance is to make the switch of narrow strips of metal with the contacts tipped with platinum, gold or alloys of various precious metals. These are then sealed into a thin glass tube and the assembly becomes a reed switch.

The switching action occurs when a small magnet approaches the reed, and when only a few millimetres away the strips of metal inside the glass tube snap together and the contact is closed.

When the reed assembly, embedded in a plastic or metal protective box, is recessed into the fixed frame of a door or window, with a magnet recessed into the moving part of the door or window, an unobtrusive security switch is achieved to sense the opening of the door or window.

What is 'normally open' or 'normally closed'?

Confusion reigns, and no wonder. If you are speaking of the device itself, such as the reed switch, 'normally' refers to the state of the contacts when not under the influence of a magnet. Therefore a reed switch intended to be used as a door contact would typically be of the 'normally open' type. When a 'normally open' reed switch is fitted into a door, and the door is closed to allow the alarm to be set, the magnet in the door changes the reed switch to 'closed' (not 'normally closed').

From the alarm circuit or overall system point of view, it is natural to think of the 'normal' situation as that in which, for instance, the system can be locked up and left for the night. The alarm signal loops would be closed, and systems consultants and users are right to think of that as being the 'normally closed' condition. If you asked for a 'normally closed' reed door contact to meet this condition you would get the wrong thing.

To avoid confusion, it is necessary to be quite sure whether one is speaking of the device itself or of the system into which it is wired.

Anti-tamper reed switches

Just as the reputation of the mechanical door switch was blemished by the chewing-gum story, so magnetic reed type contacts have had to live with the idea that they can be inhibited by a knowledgeable intruder wielding a powerful magnet outside the door. On test this is possible, but in practice the difficulty of getting the door open while maintaining the magnet in the critical position is likely to persuade a potential intruder to try something else, or to go away.

If the risk is such that the reed switch cannot be trusted on its own, firms such as Sigma supply magnetic contacts fitted with magnetic shields aimed at protecting the reed from any magnet other than that in the door frame, and with auxiliary tamper switches to warn of the presence of unwanted magnets. Examples of these are illustrated in *Figure 13.1*. Given

Figure 13.1 Maglock Securilock magnetically operated security switches: for steel doors (*left*); for non-steel and wooden doors (*centre*); actuator magnet (*right*). (Supplied by Sigma Controls Ltd)

adequately strong magnets and good installation, reed switches give good security, with reasonable tolerance to wear and warping and to vibration, which otherwise could cause false alarms.

An important point in installation of magnetic contacts in steel doors is to prevent the magnetic field from being unduly affected by the steel of the door. Typically, this is done by cutting a hole in the steel of at least twice the area of the magnet face and using non-magnetic material such as brass or aluminium to support the magnet in the door.

Door wiring

One thing an intruder might think of as an alternative to defeating a magnetic door contact is to make a hole in the door itself, large enough for him to get through, as shown in *Figure 15.1*. A defence against that possibility is door wiring. This consists of an irregular pattern of single-strand brittle wire stapled on the inside of a door at risk, and the wire then concealed and protected quite tidily by fitting a sheet of plywood or equivalent over the wiring on the door. Cutting a man-sized hole in the door will sever at least one of the loops of sensing wire concealed in the door and will cause an alarm to operate.

Tube and wiring (and foil) for windows

If potential intruders decide that it is not worth the risk of attacking a door they may well try a window. A long-established but still acceptable form of intruder detection is tube and wiring. To the uninitiated, when installed to protect windows tubing looks like steel bar physical protection. In fact, a series of vertical tubes spaced, say, 100 millimetres apart would have threaded through them a single strand of brittle wire, anchored at the ends of each tube. Unwary intruders may be quite surprised to find how weak the 'bars' are when they pull them apart to gain access to windows and equally surprised to find that they have triggered the alarm system in doing so. The alarm is triggered when the brittle wire in the tubes breaks and opens an electrical circuit connected in the alarm system. This is a crude but effective form of switch, and tube and wiring assemblies are found at least as often inside as outside the windows being protected, where they are less susceptible to tampering and corrosion. Where bars are unacceptable, an alternative to wire is foil strips glued to the glass, the circuit being broken as the glass is broken.

Pressure mats

In principle, a security pressure mat installation involves the distribution of mats within the protected area, located in positions where the intruder is likely to have to tread on at least one mat while attempting to achieve his objective. They are hardly suitable for use out of doors, so their main application is as back-up in case the perimeter of the building is penetrated. To be effective, the mats need to be concealed from view, and consequently they are used mainly where existing carpets, doormats and staircarpets offer sufficient cover.

Many construction methods have been devised, but typically a mat will consist of two sheets of metal foil separated by a thin sheet of foamed plastic, the whole contained in a flat bag of plastic sheet or moisture-resistant fabric. To function as a switch when stepped upon, the sheet of foamed plastic separating the metal foil sheets has a pattern of holes cut in it, with the hole size and spacing chosen to ensure that normally the foils are kept apart and yet are brought together to make contact under the pressure of the heel or sole of a shoe as it steps on the mat. The contact between the foils so made acts as a switch to trigger an alarm system.

Flat unobtrusive cable has been devised for interconnecting a series of pressure mats, and there is much to be said for this form of simple inexpensive trap detection. Mats are designed to avoid being triggered by the normal range of domestic animal, but if all risk of false alarms from them is to be avoided so should animals. Great care is needed in the location of the mats relative to furniture, and there have to be limitations on the subsequent movement of furniture. Mats do not trigger when, for instance, a chair is placed over them, but later they settle gradually until an alarm is triggered, say, during the night.

Of necessity, pressure mats are normally 'open-circuit' and an alarm is caused when the switch is closed. Therefore if during the day a wire to the mat is cut accidentally or maliciously no warning would be available, and the mat would be ineffective. The introduction of tamper loops and other devices has done much to minimise this hazard, and pressure mats remain a worthwhile intruder detector.

Inertia switches

One fascinating aspect of security is to see how quite well-known physical properties are put to good use in a totally unexpected way.

Principles of operation of inertia switches

When the idea of inertia was first becoming generally known, a favourite party piece was to place a cup and saucer on a tablecloth on a table. The trick was to flick the cloth away from under the cup and saucer, leaving them safely in the same position on the table. If the trick was performed successfully, it was a demonstration of inertia of the cup and saucer relative to other movement.

Now, to visualise how this idea can be applied to security, imagine, if you will, that you have a small three-legged stool and that you are holding the stool in the unlikely position of legs-up and seat-down. You also need to imagine that a colleague has found an outsize football and placed it so that it rests on the three upturned legs of the stool. If your colleague then asks you to walk slowly towards him, the chances are that the football will stay comfortably in position on the stool. Your colleague may have other ideas, and may suddenly call 'Hey, you!' from behind you. As you swing round in surprise, the chances are that the ball will topple off the stool – its inertia helped to keep it in its original position, while the stool moved sharply as you turned.

To convert the stool and ball into an inertia switch, we have to shrink the ball to, say, 10 millimetres diameter and to make it of metal. Also, the three legs of the stool have to be of metal, of such a length and spacing that they support the ball without it settling too low between them and without the ball being perched too delicately on top of the legs. Then we connect the legs to a circuit, so that if the ball loses contact with a leg the circuit is broken and an alarm signal is raised.

Applications of the principle of inertia

The potential applications of this principle are quite wide-ranging, covering virtually any situation where an intruder might cause vibration. Actual applications include the protection of filing cabinets, windows, doors, and so on, through to the very difficult job of sensing attacks on outdoor perimeter fences. As explained in Chapter 6, there are other ways than switching to reap the benefits of the principle of inertia, and for outdoor perimeters the use of mechanical inertia contacts can lead to rather too fine a balance between certainty of detection and false alarms for comfort.

Mechanically they are G- or acceleration-sensitive devices, and while they can stand movement of, say, 5 millimetres peak-to-peak at frequencies below 10 Hz without triggering, at frequencies of 1000 Hz or so, such as can be generated by moving a moistened finger across a pane of glass, an inertia switch fitted to the window would trigger immediately, minute though the vibration movement of the glass might be.

Electrically the inertia switch has only the contact area of the metal ball resting on the contacts. Contact resistance can be minimised by using gold or other precious metal plating of the contacts, but no artificial pressure can be applied other than by a small magnet to help gravity. The safe current rating of the contacts is therefore very low, and expedients of using sensors or extra contacts in parallel do not eliminate the risk of contact welding or contamination.

The various manufacturers of inertia-type switches are well aware of the problems, and it is wise always to use the maker's 'analyser' in conjunction with the switches to limit the current and to help distinguish between real and false alarms. As their name indicates, the pioneers in this field are First Inertia Switch Ltd, and their products have stood the test of time.

Mercury switches

When we tried to design a switch earlier in this chapter we soon came up against contact corrosion and pressure problems. A surprising expedient is to use mercury as a contact material. This is remarkably successful in overcoming these problems, but a disadvantage is the need to tilt the switches to make them operate. An obvious way of making good use of this 'limitation' is to fit them to windows that tilt to open. If a villain tries to get in through such a window fitted with a mercury tilt-switch our alarm could be triggered.

Out of doors mercury switches can be an alternative when high-frequency vibration of stretched wires in fences caused by wind gives trouble with other fence sensors.

Panic switches

These are more representative of an application of switching, rather than of a switching technique. Nevertheless, it seems that the correct technique for this application is not always used.

It is difficult to predict whether, in an emergency, a person under mental stress would keep a finger pressed hard on a push button for at least half a second, or whether there would be time only to give the button a momentary press.

In Chapter 15 reference is made to sampling time, and the minimum time taken by a person to move. Panic buttons are the exception to the rule. If the remainder of the system has been designed not to trigger on very brief signals as an anti-false-alarm measure then the signal from a panic button needs to be 'stretched', to ensure that, however momentarily the button is pressed, the resulting signal lasts long enough to pass the system sampling time and allow an alarm to be raised. Stretching can be chosen from such methods as a lock-in button, which needs to be reset later, a pneumatically delayed release button, or electronically. In considering the pros and cons of panic buttons decisions should turn on the view that people are even more important than the avoidance of false alarms.

Pneumatic switches

In the example of a panic switch, air can be used to delay its operation. Another use of air can be to cause the operation of a switch, as with the type of presence detector used in the forecourts of garages and petrol stations. A length of neoprene-type tubing is laid across the forecourt and when a vehicle passes over it the air pressure in the tubing is changed sufficiently to operate an air pressure valve in the attendant's kiosk. A switch attached to the air valve alerts the attendant to the arrival of another customer.

In security this type of remotely controlled switching has limited but very real applications where a risk can occur within an area that is required to be intrinsically safe. The pneumatic tube may be coupled to a floor air-pressure mat, or to a door-operated pressure valve in the risk area, thus removing entirely the possibility of an electrical spark igniting flammable vapours.

Detail design and installation has to be carried out in cooperation with the safety officer concerned, and it is interesting work if you can get it.

Hall-effect switches

Edwin Hall was born in 1855, and he discovered that if a magnet is placed close to a semiconductor the current through the semiconductor can be changed. He would have been delighted to see what we call semiconductors, but perhaps disappointed at the relatively few applications we make of his discovery.

With a Hall-effect device the durability and design worries of mechanical contacts are avoided. Current ratings available are comfortably within the range that can be used reliably in security electronics, and the device has properties that make it that much more difficult for the villain to tamper with.

Their very high reliability makes them particularly suitable for use where inaccessibility would make servicing of, say, microswitches difficult and expensive. The ratio of first cost to service cost is already heavily in its favour, but that is still a hard concept to sell. They do need a power supply, so in application they have to be thought of like any other electronic sensor.

Touch switches

To complement the magnetic proximity switching effect of the Hall switch is the capacitive proximity effect of the touch switch used in passenger lifts in buildings and in various types of keyboard. Its principal application in security is where resistance to vandalism and mechanical damage is important.

Discussion points

The choice of type of switch is wide enough to meet most applications, and selection is likely to continue to follow a fairly consistent pattern to suit. What often turns out to be tiresome in the field is the provision made for fixing the switch to the structure; or the method used for attaching the connecting wires; or its shape, which leads to unnecessary work in preparing a recess to conceal it in the structure. It is worth gathering a few sample switches and inviting a design engineer and an installation engineer to join you in looking at switches from the field point of view. The resulting discussion should be quite revealing.

14 Active infrared sensors

It may not be enough, or practicable, to use any of the range of switches described in Chapter 13 to detect intruders in a particular situation, and one of the earliest alternatives or additions to the switch was the use of a beam of light aligned across a route likely to be taken by an intruder. The beam would be produced by an electric light bulb with a lens in front, as in a hand-torch, all forming a transmitter. A photoelectric receiver would be aligned behind a lens at the other end of the beam. If an intruder walked through the beam, momentarily cutting off the light reaching the receiver, an alarm could be raised.

In the seesaw war of measure and countermeasure beams of visible light soon became useless as villains realised what they were. The next stage in development was to make the beam invisible, by covering the transmitter lens with a filter. Because it so happened that much of the energy already being transmitted by the light bulb was in the invisible infrared region, that became the pass-band for the filter chosen.

Beam systems have the advantage of 'failing-safe' – that is, an alarm is raised if the transmitter ceases to function owing to light-bulb failure. Inevitably this happened rather too often, owing to the long operating hours necessary. The problem of bulb failure was solved by replacing the filament bulbs with semiconductor light-emitting diode-type infrared sources when they became available. The resulting infrared beam sensor has established a seemingly permanent place in the intruder-detection armoury, for both indoor and outdoor applications, and examples of typical equipment are described below. Passive infrared systems, which do not need a light source, are described in Chapter 17.

Active infrared equipment

Some equipment is claimed to be suitable both for indoor and outdoor operation, but cost and environmental requirements lead to separate versions, and probably that is the best way of looking at them.

Indoor active infrared sensors

In learning about active infrared beams one is bound to hear the attendant folklore, such as the use by an intruder of military-type infrared binoculars to 'see' where the beam is, and the possibility of holding off the receiver trigger by shining another transmitter at the receiver while the intruder breaks the beam. Although such techniques are feasible, the intruder is unlikely to bother with them. However, just in case, two things can be done. First, the lens used on the receiver can be designed to accept infrared light over a rather narrower angle than originally thought reasonable. More importantly, the light-emitting diode transmitter can be modulated, or made to flash on and off at any of a very large range of frequencies. If it is thought that the intruder can find out the frequency chosen, and can make up and use equipment to simulate the transmitter, then the protected risk must be worth a great deal to the intruder, and additional or alternative detection devices are justified.

Therefore for anything up to moderate risks the folklore can be ignored. That one can survive by making simple things well is illustrated by Radiovisor, a British company that first used invisible beams in 1929 to protect a collection of silver during an exhibition. Now known as RV Limited, their type M125 general purpose model is shown in *Figure 14.1*.

The light-emitting diode transmits energy at a typical wavelength of 940 nm towards a receiver lens having a fairly narrow acceptance angle of 5°. The neatly styled housing conceals pivots which allow both the transmitter and receiver to be adjusted horizontally through 180°. However, the angle used is concealed from the intruder by the fully radiused black-looking combined filter and cover. The beam is modulated, with a range up to 125 metres.

Figure 14.1 Active infrared beam intruder detector Type M125. (Supplied by RV Ltd)

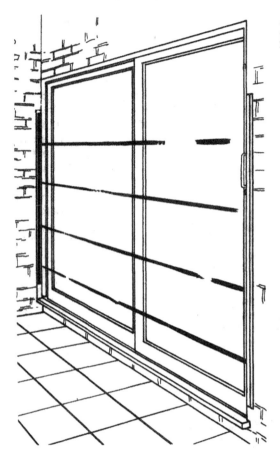

Figure 14.2 Window intruder detector. Patent No GB 2203239B (Supplied by RV Ltd)

A more recent addition to the AV range is the 'Wingard' multi-beam barrier for mounting inside windows. This goes a long way towards solving the dilemma of risking undetected entry or of risking false alarms from most window sensors.

An American alternative to the RV long range active beam of *Figure 14.1* is offered by Pulnix, and their type PR-5B is shown in *Figure 14.3*. The attractive and rather unusual feature of this active infrared sensor is that the transmitter and receiver are contained in the same housing. The transmitted beam is reflected back by a mirror placed where the receiver would normally be. The advantage is, as with passive infrared and with ultrasonic and microwave detectors, that only one unit has to be wired, compared with the two units (separate transmitter and receiver) for the conventional active infrared system. However, systems designers and installers have to remember the optical law learned in physics, that 'the angle of reflection is twice the angle of rotation'. In other words, if the reflector is moved 1° out of true alignment the reflected beam is moved 2° out of alignment, which can be a significant proportion of the safety factor built into the system. Pulnix recognise this by limiting the recommended beam length to 5 metres in the type PR-5B, and to 10 metres in another model, ample for many indoor applications.

Active infrared for outdoor applications

There is much more in favour of using infrared beams out of doors, and Chapter 6 covered some of the more important points. As they are inherently line of sight devices, they are not too good at following ground undulations and changes of perimeter direction, but they take up very little ground area, and, compared with some alternatives, they are relatively trouble-free. Evasion is feasible; and if it matters for the risk involved, they can fail to operate during thick fog. But they keep going long after visible light has been cut off, simply because the wavelength of infrared light is longer and the energy suffers less attenuation and loss by scattering when penetrating the moist fog droplets. Typically, the wavelength is about 10 μm (one-thousandth of a millimetre) which is about 20 times the wavelength of visible light (0.5 μm for green).

Outdoor applications range from a single beam across a roadway to detect vehicle movement through to barrier beams consisting of three or more beams stacked one above the other to minimise evasion of detection by an intruder, either along a passageway or through a perimeter. The usual test of what you would do if you were a villain (Chapter 2) shows how and how not to use infrared beams outdoors.

For instance, they can be used for perimeter detection when sited between a boundary fence and an inner security fence. They are particularly useful inside a not-too-substantial combined boundary and security fence of chain link type, but that proviso leads us to see that infrared beams should not be used indiscriminately just inside a substantial brick wall, because the strength of the wall would make it easy for an intruder to jump clear of detection by the infrared beams.

When multiple beams are used, there is scope for various combinations of beams, depending upon the emphasis given to freedom from false alarms relative to certainty of detection. The duration of interception or interruption of a beam can also help, on the basis that it takes a person a known finite time to go through, and anything taking less time cannot be a human being.

Figure 14.3 The Pulnix Type PR-5B active infrared sensor, which uses a reflector to return the transmitted beam to a receiver built into the same housing with the transmitter. (Supplied by Pulnix America Inc.)

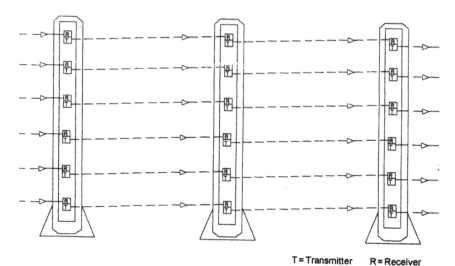

T = Transmitter   R = Receiver

Figure 14.4   Multiple active infrared beams can be arranged in towers 1.8 metres high, spaced up to 150 metres apart, when using Rayonet 2000 equipment for outdoor perimeter detection (Supplied by Integrated Design Ltd)

A leading exponent of the use of multiple active beams outdoors is Integrated Design Ltd, and an example of how their Rayonet 2000 is used is illustrated in *Figure 14.4*. Each tower contains six transmitter and six receiver units, to create six parallel beam zones. Intruders find it hard to evade such a system, so the main problem is avoidance of nuisance alarms. Logical thinking has enabled electronics to be designed to make the decision from information given by a beam break pattern that 'cannot have been caused by a human being', and therefore no alarm is raised. It is easy to understand this when, for example, a bird flies through a beam. It would take a man longer to get through a beam, and that time difference can be used to prevent a false alarm.

Discussion points

Bearing in mind that after door contacts and pressure mats, active infrared sensors were one of the earliest intruder-detection devices, and are still in regular use, how would you prepare a risk analysis to prove your case for specifying the active infrared method in a particular situation? Can it be done factually or would you have to rely on precedent and persuasive opinion?

15 Ultrasonic sensors

Why ultrasonics?

There was a time when the typical intruder detection devices available were door contacts, pressure mats and point-to-point infrared beams, as described in Chapters 13 and 14. If these devices were evaded, there was no way of detecting the intruder moving around in the building.

Figure 15.1 illustrates the point. By the standards of the day the building was well equipped with intruder detection devices, but the intruders had apparently found out that by cutting a hole in the door they could get in and out again without being detected. Even if the doors had been 'wired', another weak spot in the perimeter or roof might have been found, and better security measures had to be developed. The pioneers found themselves concentrating upon the two things that can obviously permeate the sort of space encountered in buildings, radio and sound. In Chapter 16 we shall see why radio had to come later.

Figure 15.1 Penetration of an art gallery. Detectives examining the panel in the side door of the gallery where the thieves made their entry. (Supplied by *The Times*)

In anything to do with space detection it helps to use analogies, and this we shall do in this chapter as we did in Chapter 4. Also, it helps in this particular case to use events in the historical evolution of space detection to show how problems were met and overcome.

Audible sound detection

The use of audible sound was and still is practicable if a security officer is on the premises to recognise the cause of most of the sounds he hears via the microphones, but it was not practicable for a security officer based at a central security monitoring station to recognise and deal with all the sounds coming to him from buildings all over the town or area, nor could he listen to more than a very few at a time. Later developments went some way to overcoming these difficulties, and Chapter 18 is included to cover audio sensors.

What else was there?

Listening directly to sound on a large scale was therefore out, but using sound had to be in, really because there was nothing else. This chapter is concerned with why ultrasonic sound was decided upon, how initial troubles were overcome, where to use it successfully and what can cause nuisance alarms.

Pioneering thinking

As a first step the pioneers decided that instead of relying upon hearing whatever sounds happened to be audible in the buildings, they would create their own sounds which would need no interpretation, and to arrange matters so that the security officer need give no attention to them except when an intrusion occurred. Such a system clearly would still need a microphone, which would be just as prone as before to picking up any sounds that were there to be heard, man-made and natural. The only way they saw of getting round this was to make their own sound work at frequencies outside the normal range of human hearing, and to make the microphone and receiver ignore audible sound.

They could have gone to frequencies below the human hearing limit (subaudio frequencies) but not enough was known about this region, and electronic techniques were not very well developed for subaudio operation. Much more was known about frequencies above the human hearing limit (ultrasonics) through the work of the pioneers in acoustics, and electronic techniques were readily available for this new application. The decision, then, in favour of ultrasonics was arrived at more by a process of elimination than by direct choice – a feature, as we will see, that is quite common in security engineering.

The properties of sound

Two quite different ways of using ultrasonics were developed: first, the standing wave system and later the radar Doppler system. To understand these it is necessary to know more about sound itself and the part played by the air. If you have not yet studied Chapter 4, on the fundamentals of space detection, please do so now, and then come back to this point.

The standing wave system

Although the standing wave method became unpopular when used with ultrasonics, it is important to understand the basic principles of how it works before dealing with ultrasonic radar systems.

Volumetric coverage

For the standing wave system the technique adopted involved fixing a transmitter of ultrasonic sound fairly high up on a wall of the room, and an ultrasonic sound receiver similarly high up on the opposite wall. The type of transducer used for transmitting was chosen to give uniform energy output in all directions as far as was possible, which usually meant 180° horizontally and 45° or more vertically. The receiver transducer had similar coverage. Setting these units well above floor level ensured that they were reasonably free of obstructions and approached the ideal of some ultrasonic energy travelling direct from the transmitter to the receiver.

However, in addition to some direct reception, the receiver will also pick up energy reflected from all directions – from the walls, floor, ceiling and anything within the room. The receiver converts the acoustic energy into electrical energy ready for use in the electronic signal processing circuits. If all is stationary in the room, all the reflected energy combines with the direct energy to produce a steady electrical output from the receiver transducer. Remember from Chapter 4 that the transmitted energy is in the form of acoustic waves, or successive rises and falls in air pressure at a rate corresponding to the transmitted frequency.

When the air and all else in the room is still, it is not difficult to visualise that the ultrasonic energy settles into a steady or standing wave pattern. The importance of this condition is emphasised by the opposite situation, when an intruder gets into the room by, for instance, cutting a hole in the door. Some of the ultrasonic energy that was reflected from the stationary door panel is now being reflected from the moving intruder. Thus, the standing wave pattern as sensed by the receiver transducer is upset, and the change is converted into an electrical signal which can be used to alert a security officer locally or elsewhere.

What are the vital things we have said so far regarding the standing wave method?

The transmitter and receiver transducers are omnidirectional, to give volumetric cover against an intruder.

The received signal is made up of energy reflected from all over the room, together with some energy received direct from the transmitter.

The air and everything in the room except the intruder is still.

Good. We have a detection system which will work even if the intruder succeeds in evading all the perimeter detection devices; consequently, it was widely used.

Problems in design and use

Unfortunately, it was used also in situations where one at least of these requirements was not satisfied, and these installations gave both the suppliers and the users so much trouble with nuisance alarms that a better way was needed. Engineers regard it as good procedure to investigate the performance of existing equipment before going on to develop new equipment. By doing that with the standing wave ultrasonic technique, the problems soon became evident.

Resetting

It was said that it was not possible always to reset some standing wave detectors after the contents of a room had been changed or rearranged. This seemed an easy problem to solve. For the typical system, using one receiver, the problem was very similar to a problem that radio communication people have always had with fading signals and dead spots. You, too, will have had the same experience if you have listened to short-wave or long-range medium-wave radio stations. To overcome fading, communication engineers put up one or more additional aerials in different but closely related areas, and combined the signals from each in a receiver, so that if nothing was coming from one aerial, at least something useful would be coming from the others.

Relating this back to the ultrasonic standing wave detector, one can visualise that with multiple reflections from a room and its contents it could happen that one set of reflections could conceivably be just cancelled out by another set of reflections so that the receiver got no signal. Consequently, it would go into an alarm condition and could not be reset. Clearly, this problem could be solved, as in the above communications problem, by adding more aerials. But worse was to come.

Air movement

A further problem with ultrasonic standing wave detection systems was that they were liable to go into an alarm condition if the room was draughty, or if there was central or local heating in operation.

To get a mental impression of how this happens, we can simplify – oversimplify, perhaps – the effect of the air movement. First, imagine that the standing wave transmitter sends out energy towards the receiver on the opposite wall. Think of one bit of the energy as a pellet if you will. In still air the pellet will travel across the room at the same speed as any of the reflected energy. If, however, a draught develops in the room, flowing from the transmitter towards the receiver, the pellet will travel faster than some of the reflected energy. As before, the standing wave pattern will be upset and an alarm condition will be caused.

There was no way round that one. At least, one is forced to say with the Cornish folk when asked the way to somewhere, 'Well, I wouldn't start from here'.

Radar techniques as an alternative

The great thing about radar is that both the transmitter and the receiver are used side by side. Both face in the same direction, instead of being separated and facing each other on opposite walls of a room, as for the standing wave ultrasonic detection method we have discussed so far.

Returning to the pellet concept, we can see the effect of using the radar method. The transmitter sends the pellet down the room, where it is reflected from the opposite wall back to the receiver beside the transmitter. If a draught develops down the room, as for the standing wave method, the pellet will go faster with the draught, but on the return, reflected journey it will be going against the air flow, and slower. Adding the journey times together for go and return, the total time is the same as in still air, and within usable limits the receiver cannot tell the difference between still and moving air. So, even if there is a draught, an ultrasonic detector working on the radar principle should not give a nuisance alarm, because of the cancellation effect.

This prospect was good enough to establish the radar method as right in principle. All that remained was to find the right way of applying the principle.

Properties of ultrasonic radar sensors

The radar principle having been found to show promise as a means of reducing the risk of nuisance alarms due to air movement, are there any other favourable features and are there any disadvantages?

Reflected energy

The first thing to notice is that a radar sensor is in no way dependent upon energy reflected from the walls or anything stationary in the room. It is not dependent upon a standing wave pattern.

Line of sight

It is, however, dependent upon reflections, or echoes, from a moving intruder. For this it needs a clear line of sight between the sensor and the intruder, so that energy can be transmitted to the intruder and reflected back with no obstructions in the way.

Doppler frequency shift

An ultrasonic radar depends upon sensing the Doppler shift of frequency caused by motion within the room, as described in Chapter 4.

Inadvertent reflections

Unless the room or volume covered is very large, it is likely that multiple reflections will be set up inadvertently by the radar sensor, just as in the standing wave system. It is necessary, therefore, in the design of a radar sensor to ensure that the receiver responds to the presence of a Doppler signal and not to the absence or significant change in amplitude of a signal, as is needed in a standing wave receiver.

Focusing

Because we specifically do not want a standing wave pattern, we can take advantage of the properties of ultrasonic transducers, as explained in more detail below. Like light, ultrasonic energy can be focused and, with correct design and choice of transducer, a beam of ultrasonic energy as wide or narrow as we wish can be obtained.

Controlled space detection

The ability to focus the energy into a beam makes it possible to direct the energy into areas we want to cover against an intruder, and to direct the energy away from possible causes of nuisance alarm. This feature led to ultrasonic radar sensors being called controlled space detectors, as distinct from the relatively uncontrolled volumetric standing wave sensors.

Limits to coverage

The concept of forming a beam of energy suggests that the detection capability extends as far in distance as the beam is effective. The effectiveness of the beam depends upon sufficient energy being in the reflected signal or echo to be detected at the receiver. The effectiveness of the beam is also dependent upon continued transmission in air. Effectiveness ceases when the beam reaches a solid face such as a wall, door, window, packing-case, floor or ceiling. At such faces some of the energy is absorbed, some is scattered and some is reflected back towards the receiver. No significant amount of ultrasonic energy can penetrate these barriers. So long as there is no relative movement between any of these faces and the ultrasonic sensor, no Doppler signal is generated and no alarm can be caused by movement on the other side of the faces.

Short-range movement

In Chapter 4 on space detection fundamentals, attention was drawn to the rapid increase in detection sensitivity the nearer a moving object was to the radar transmitter-receiver sensor. The degree to which this matters is dependent upon the size of the moving object. Awareness of the problem has led to design techniques which go far towards eliminating nuisance alarms from close-range movements such as these. This is enlarged upon below, under 'Designing to avoid nuisance alarms'.

Natural and man-made sounds

An ultrasonic receiver is liable to pick up energy other than that emitted by the transmitter. By suitable choice of transmitted frequency beam shape and sensor location, the risk of trouble from these sources can be made almost negligible.

Licence

No Government licence is needed to operate an ultrasonic sensor.

Such, then, are the main properties of ultrasonic radar sensors. Fortunately, most of them are favourable and the remainder lend themselves to adequate control.

Achieving control

We shall look now in more detail at how adequate control can be achieved and how to make the best use of the favourable features of ultrasonic radar sensors, explaining design features as we go.

Properties of an intruder

If we use the radar principle for detection, we look for the Doppler frequency that appears in the receiver due to the intruder moving in the area covered by the sensor. A skilled intruder will know that it is technically difficult to detect very slow movement, because a bottom limit is usually set on the lowest Doppler frequency to be detected, in order to cut out spurious detection of signals from other sources. A fairly stiff limit is to say that we must detect an intruder moving 1 metre in 3 minutes, or 10 metres in half an hour. Ten metres is a typical distance we can force an intruder to travel under detection; and given that both entry and exit are necessary to complete a mission, the mission would take at least an hour to complete – a severe strain on anyone's nerve. The greater the strain the greater the chance of an involuntary movement during the hour of the head or an arm, or even a leg. Such movements would generate Doppler frequencies much higher than those that come from a very slowly moving body, and the chance of detection would be correspondingly higher.

For the less skilled, uninformed intruder, the quick dash in and out may be expected. The Olympic record speed is of the order of 10 metres per second. No intruder can match that indoors, so the information is available in conjunction with limb speeds, for setting the upper frequency limit for detection, which can be used for eliminating a further range of nuisance alarm causes.

Evasion

A question sometimes asked regarding Doppler detection arises from the fact that, fundamentally, motion towards or away from the sensor is needed in order to achieve detection. What happens if the intruder moves across, at right angles to the sensors? Fortunately, this is impossible. For it to be possible, the body would have to move in a circular path, the same distance always from the sensors. Even if the intruder were able to get into position undetected to do this, would the objective be achieved if movement were restricted to a circle? Legs and arms would also have to follow a circular motion, each motion of the correct radius, which would make the prospect of being undetected still more remote.

Sampling time

A rather more serious risk of evasion by a skilled intruder arises from shuffling, i.e. by moving a little and pausing quite still for a few moments before moving again. By moving in this way much less mental and physical strain is felt by the intruder. But, by understanding this further property of an intruder and by using sampling time in the correct way, the risk of evasion is again catered for adequately.

Extensive experimental work with a wide range of people shows that no one can move significantly in less than a certain time. Thus, we can say with fair certainty that a signal lasting less than this time when detected at the sensor cannot be caused by an intruder. Therefore, all signals lasting for a shorter time than this, such as clicks and spikes on the mains power supply, can be rejected electronically and nuisance alarms avoided. If the signal lasts for this critical time or more, there could be an intruder and the alarm has to be raised. This technique is known as using sampling time rejection, and it needs quite clear thinking in the light of the above derivation to make sure that it is not confused with purely delayed triggering. For the less skilled intruder the sampling time can be longer, and it is a matter for judgment what figure to use. On test, the effect is shown by triggering on the first movement, corresponding to at least one-tenth of a second, or after one or two steps have been taken, corresponding to, say, one second.

Frequencies usable

The higher the operating frequency chosen for an ultrasonic radar sensor the greater the prospect of avoiding nuisance alarms caused by natural and man-made sounds and their overtones. The lowest usable frequency is of the order of 20 000 Hz, and 40 000 Hz is typical. How practicable is, say, 80 000 Hz? Factors influencing the choice of frequency at the design stage include the following.

Attenuation   In Chapter 4 attenuation in air was given as roughly inversely proportional to the frequency squared. To give a scale of values, doubling the frequency from 20 kHz to 40 kHz while retaining the same beam angles and distances reduced the signal to a quarter of that at 20 kHz. Increasing to 80 kHz would reduce it to a quarter of that again or to one-sixteenth of that at 20 kHz. Clearly, any increase above this frequency can introduce problems in transmitter power generation and in achieving adequate receiver sensitivity.

Beam shape   We saw in Chapter 4 that ultrasonic energy follows very closely the physical aperture laws that apply to light and to microwave energy. Typical figures for beam widths are given in the section below on types of transducer. Again, however, increasing the frequency too far makes the beam excessively narrow and pencil-like, which is unsuitable even for controlled space detection.

Mutual interference   Given a fair compromise in choice of frequency, and given the need to put sensors in a fairly large room, it may not be practicable to cover the whole risk with one sensor. If two are used on the same nominal frequency, there is a possibility that the frequencies will drift a little and the difference will appear to one or other of the sensors as a Doppler frequency and thus cause an alarm. The safest course is for each ultrasonic sensor in any one room to operate deliberately on separate frequencies, the separation being substantially more than the maximum Doppler frequency the sensor will accept. Why not remove the problem by using crystal control to ensure that each transmitter is exactly on the same frequency? Sadly, if you think about it, you are back then with the standing wave problem. If you are a designer, you have the information needed to do the right thing. If you are a customer, this information will guide you on what questions to ask potential suppliers of equipment.

Types of transducer

To use the definition given in Chapter 4, an ultrasonic sensor is an active device, meaning that a man-made transmitter of ultrasonic energy is needed. Electrical energy is converted to acoustic energy by the transmitter transducer. The Hi Fi loudspeaker is a good example of the principle. This leads us into the simplest of the transducer types used.

We saw in Chapter 4 how the beam shape is controlled by the shape and dimensions of the transducer. In a Hi Fi loudspeaker, if the beam width is to be kept wide enough to preserve the stereo stage, the loudspeaker diaphragm has to be the smaller in diameter the higher the frequency the speaker is to radiate satisfactorily. Ultrasonics needs still higher frequencies, so the diaphragm has to be still smaller in diameter to keep the beam width large enough.

Disc   A typical ultrasonic sensor transducer is illustrated in Figure 15.2 together with a two-dimensional attempt at showing its three-dimensional coverage pattern. In this instance the transducer is a flat disc of material of barium titanate type, of about 10 mm diameter. This material is used in some Hi Fi tweeter loudspeakers; but whereas for the speaker application we need a wide-ranging high-frequency response, for the ultrasonic application we

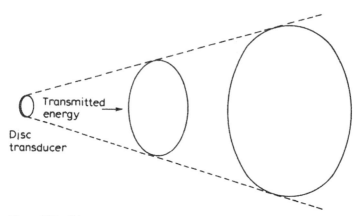

Figure 15.2   Disc transducer and conical beam coverage pattern

need sensitivity at one frequency only, and the piezoelectric material can be designed to resonate at this required frequency. To complete the arrangement for an ultrasonic motion sensor, we need also a receiver transducer, giving the same coverage pattern as the transmitter. By placing the transmitter and receiver side by side and facing the same way, the coverage patterns can be made to overlap, as shown in Figure 15.3.

The beam shape from a disc transducer is quite sharp, typically 40°, and while this is an advantage in siting to help avoid looking at something that could cause a nuisance alarm, it can make it easier for an intruder to evade detection.

Ring   A second type of transducer is illustrated in Figure 15.4. It consists of a fairly thick ring, again of barium titanate piezoelectric material. In use the axis of the ring is usually vertical, with the transmitter ring mounted 300 millimetres or so above the receiver ring. Radiation from the ring gives 360° all-round coverage, but this is usually restricted to 180° on a wall or to 90° in a corner. In the vertical direction the coverage is controlled by the height chosen for the ring, but 60° would give very fair coverage against evasion in most situations.

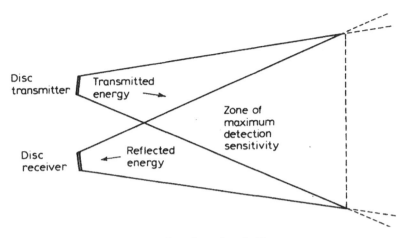

Figure 15.3 Sensitivity pattern of overlapped conical beams

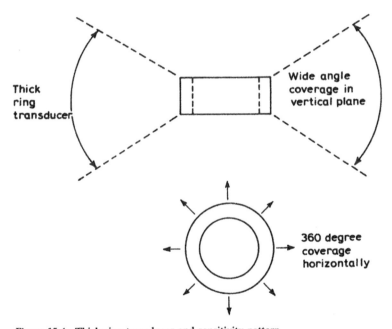

Figure 15.4 Thick ring transducer and sensitivity pattern

Tube The third type of transducer is in the form of a tube, and it should be possible from what has been said above and in Chapter 4 to work out what coverage pattern to expect. The narrow pattern in the vertical plane can be made still narrower if the ultrasonic frequency is increased. With frequencies between 40 and 80 kHz, this is the purpose of transducers of this shape – to give coverage of 90–360° in one plane and only 10–20° in the other.

Considerably greater flexibility and total effective range, control can be achieved by using the location of looking downwards from a ceiling or roof-truss. This aspect is discussed a little later in this chapter.

Designing to avoid nuisance alarms

Many of the causes and ways of avoiding nuisance or false alarms are referred to in Chapter 4, in this chapter, and elsewhere. The following is therefore more in the nature of a quick reference check-list, with relevant additions.

Understanding

The designer, above all, needs to know and understand the fundamental properties of air, transducers and electronics, and reference to appropriate textbooks is necessary for more advanced treatment.

Experience

You don't get a 'feel' for a subject just by reading. Check every point you can by noticing what happens in life, set up experiments to get a sense of values – what happens to this? – in relation to that?

Reasoning

Combine experience with understanding to find a better way. Again it means considering the dual question 'Will it work?' 'What can go, or be made to go, wrong?' in the hostile environment in which security equipment has to survive.

Total cost

The total cost of a product is the sum of the purchase price, operating cost and maintenance cost. Commercial pressures concentrate on reducing the selling/purchase price, at the expense of the other two. Designers usually want to offer quality and reliability. Can you give your sales people enough selling points and conviction to sell quality at a price perhaps higher than that of the competition? Can you find a way of offering quality at prices below those of the competition? Can you devise a product that has no competition?

Information

Good design can be lost on the purchaser, on the installer and on the user unless specific efforts are made to get key information through to them, through personal training, leaflets handed over as part of the product and installation, and, perhaps, informative labels inside equipment covers. Incidentally, it is important to give interconnection diagrams to emphasise the correct use of the equipment in relation to the rest of the system.

Coverage patterns

There is really no reason why transducers should not be designed to give the optimum patterns needed for detection and nuisance alarm avoidance. Too often transducers designed for other jobs such as remote control are used. Ultrasonics obeys the optical aperture laws, and coverage patterns are easily manipulated with correct transducer design and in some cases the use of reflectors.

Transducer frequencies

To avoid man-made interference, frequencies of 40 kHz and over are needed. To avoid mutual interference between sensors in the same room, several fixed frequencies at non-harmonically related intervals are needed. It is not enough normally to rely on crystal-controlled common frequencies fed to all transmitters. This method can nullify the advantages of using the radar principle for avoiding nuisance alarms in draughts and air movement generally, because the transmitter in one sensor can work with the receiver in another, akin to the troublesome standing wave arrangement.

Transducer separation

Whatever its size, the closer a moving object is to the transducers the greater the prospect of an alarm. The problem of moths and flies can be made negligible by separating the transducers. It follows that if a moth is not 'illuminated' by the energy radiated by the transmitter transducer, there can be no echo from it to the receiver, and vice versa. The separation needed is determined geometrically from the transducer beam shapes used for the coverage pattern required. (See also Chapter 16.)

Standing wave–Doppler separation

Is phase-sensitive detection really necessary? It makes ultrasonic Doppler systems 'touchy', to say the least. Tests with a skilled slow mover show that he makes quick limb movements involuntarily from time to time, which can give signals comfortably within a pure Doppler pass-band.

Sampling time

The skilled intruder needs 100 ms or more for any movement. Anything lasting for a time less than this cannot be an intruder and can be filtered out of the alarm signalling circuit, and with it will go a variety of click-type causes of nuisance alarms.

Earthing/grounding

It is customary to earth electrical equipment either via the neutral line or directly via a separate earth wire. Similarly, thermionic valve electronic equipment is earthed. Why not semiconductor equipment? There is a new-found awareness of the need for earthing, but all too often it is omitted in completed installations of security equipment. Attention to details such as this can get rid of baffling nuisance alarms that just refuse to occur when you are on site to investigate.

Range control

Detection sensitivity should be no more than needed for the distance to be covered in each location of an ultrasonic sensor. In order to avoid inadvertent or deliberate turning of the range control to zero, it is good practice to limit the range control to not less than 25% of full range. Setting the range correctly on site needs a control that has been designed to allow for the radar inverse fourth-power law discussed in Chapter 4.

Signal-to-noise ratio

It is worth devoting considerable effort to minimising the receiver noise and to maximising its stability. A sure sign of latent trouble is when the auto-reset alarm output from a sensor 'hangs on' after a test movement has stopped, instead of resetting cleanly after a very few seconds.

Test equipment

A walk test has to provide the final criterion, but on site it can be time-consuming if it is the only regular means of setting up. A hand-held receiver or field strength meter gives positive indication that transmitted ultrasonic energy is getting to wanted places and not getting to areas that could cause nuisance alarms.

Suitable applications

In considering suitable applications of ultrasonic sensors it is assumed that the sensors are of the Doppler radar type, conforming to the optimum design, generally as given above.

Glass

The outstanding and almost uniquely suitable application of ultrasonic Doppler sensors is in risk areas containing glass. Ultrasonic energy does not penetrate glass significantly, which avoids nuisance alarms from legitimate movement outside the risk area. Nor is there much outside the risk area that can penetrate the glass to affect ultrasonic equipment inside the area.

Large areas

Given that the designer has taken the opportunity of freedom from Government control of transmitter frequency, ultrasonic sensors can be available for operation on several different frequencies, which allows a number of sensors to be used to cover a fairly large area without mutual interference. The total number of sensors can be controlled reasonably well by siting them in regions where the intruder would most likely have to be to achieve his objectives.

Small areas

Still using the word 'area' in its colloquial sense of enclosed volume, ultrasonic sensors lend themselves particularly well to small high- and low-risk areas which may be totally enclosed.

Open areas indoors

It is customary for ultrasonics to be used as a last link in a chain, to detect an intruder in the risk area after he has penetrated physical and electronic perimeter defences. But the sensitive detection zone of ultrasonic sensors can so readily be controlled that a good application is to the approaches to a risk area, giving warning before the risk area has been penetrated.

Contrary to the impression the above list may give, ultrasonic detection is no panacea. It is subject, like other forms of detection, to restrictions arising from the causes of nuisance alarms. The aim of these chapters is to bring together the surveyor and the engineer in understanding how to choose the best way, be it ultrasonic, microwave, infrared, or what have you.

Applications to avoid

All too often the selection of the most suitable detection method in any situation is more a matter of finding what is left after unsuitable methods have been eliminated owing to environmental problems. It still happens that just nothing seems suitable, so there is scope for new ideas.

There is little in the way of intruder movement that an ultrasonic Doppler sensor cannot detect; so, in checking which applications to avoid, we can concentrate on the factors that can cause nuisance alarms. Drawing upon the earlier remarks on causes of nuisance alarm, we have the following factors.

Outdoors

An ultrasonic radar Doppler sensor can detect as effectively outdoors as indoors, but the risk of nuisance alarms from natural causes is increased so much that it is normal to avoid outdoor applications. The number of successful outdoor applications will increase as designers and systems engineers succeed in avoiding nuisance alarms from rain, wind-blown objects, high-velocity winds, birds and animals.

Rotating blades

Whereas ultrasonic Doppler radar sensors are inherently resistant to nuisance alarms from reasonable air movement, they can be sensitive to the rotation of fan-blades and the like, caused by air movement. Ultrasonic sensors are more likely than other space devices to detect blades when motor-driven at normal speed owing to an unfortunate chance relationship between blade speed and the sensor ultrasonic frequency/wavelength characteristic giving Doppler frequencies similar to those we expect from an intruder. Fan-blades some way down an open-ended ventilating duct can be just as troublesome.

Wavering

Movement of an intruder towards a Doppler sensor causes a reflected signal apparently higher in frequency than the transmitted signal. Similarly, movement away causes a lower Doppler frequency. A curtain or blind may waver to and fro in a draught, producing a succession of Doppler signals above and below the transmitted frequency. It may sound technically easy to use this principle to cancel out wavering signals, and indeed sensors are available which incorporate this feature. But care is needed in using cancellation if one of the strong features of ultrasonics is to be retained – the ability to detect very slowly moving intruders. So, in cases where wavering objects cannot be avoided, it is preferable to decide whether the risk of not detecting a skilled intruder is more acceptable than reducing the nuisance alarm risk.

Vibration

The resonant frequencies of buildings tend to be below the range of frequencies normally accepted by ultrasonic Doppler radar sensors. An example of a typical exception is the rattle of a window caused perhaps by resonance with a vehicle engine outside. An unexpected example met in a bank vault in London was caused by underground trains. One would expect any such vibration to be outside the Doppler range of frequencies, but on investigation it was found that the vault was extremely stiff, making extensive use of steel reinforcing and of concrete that had been concentrated by deliberate vibration before setting. Trains caused the vault to vibrate at around 70 Hz, unavoidably within the Doppler range required. At that time no acceptable alternative detection method was available, and the sensors had to be mounted on specially designed antivibration mounts. Short of such measures, ultrasonic sensors should be avoided in situations where the risk of nuisance alarms from vibration cannot be avoided by suitable siting.

Telephones

In the UK the insistent bell sound of the older systems contained overtones extending well into the ultrasonic region, causing considerable trouble with ultrasonic sensors. The current, more melodious American-style telephone sounders generate fewer overtones and cause less trouble.

Air lines and steam pipes

Just as an audible note is produced when a wind musical instrument is blown, so an inaudible ultrasonic note can be produced when an air or steam pipe under pressure develops a leak. The frequency of the note may cause interference with ultrasonic sensors. Unless adequate means are available for tracing and repairing leaks of this kind, ultrasonic detection should be avoided near air lines and steam pipes.

Limits of air movement

When radar techniques were being considered for ultrasonic detection, it was noted that the sum of the journey times of a transmitted signal and of the reflected received signal was approximately the same in draughts as in still air, and that within usable limits a radar system could operate satisfactorily in draughty conditions. This applies equally to convection air currents from heated radiators and the like. Beyond limits that can be designed into the sensor the faster the air moves the more the risk of false alarms increases, and there is no point in asking for trouble. Good and careful siting of the sensor is the key. If you cannot find a better location for the sensor than directly over a radiator, for instance, don't use ultrasonics.

Dual technology

Such are the attractions, however, of ultrasonics that some are reluctant to take this advice and they seek ways round it. Thus the dual-technology concept has emerged, in which, say, an ultrasonic sensor is combined with a microwave radar sensor and connected so that both must trigger simultaneously to confirm an intruder situation. If, however, the ultrasonic sensor triggers in hot air movement the microwave is unlikely to, so a false alarm can be avoided. The dual-technology concept is looked at in more detail in Chapter 19.

Proprietary equipment

There are now so many manufacturers of equipment that it would be an invidious task to select from them what might be interpreted as 'recommended' versions. As with other proprietary equipment mentioned, the ultrasonic systems selected are typical implementations of the techniques discussed, or are products with which I have had contact in either a design or user capacity.

For the sake of historical completeness the pioneering ultrasonic Doppler radar developed by Decca Radar Ltd and sold for several years under the name Deccalarm is illustrated in *Figure 15.5*. Features of interest include what seems now to be its rather large size, which stemmed from a 'no compromise' design to keep transmitter and receiver transducers sufficiently far apart to eliminate the risk of false alarms from nearby disturbances such as moths. Also of interest, perhaps, is that the Deccalarm shown was designed from the outset for ceiling or roof-truss mounting, which gave the maximum practicable floor coverage area. Floor coverage was increased yet further by using reflectors set at an angle

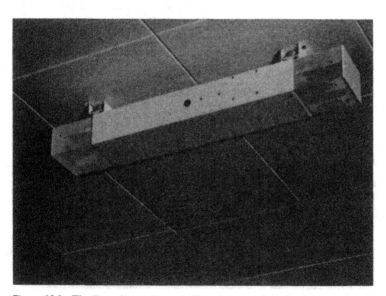

Figure 15.5   The Deccalarm ultrasonic Doppler radar intruder detector developed by Decca Radar Ltd, shown here mounted on a ceiling. (Supplied by Decca Radar Ltd, now part of the Racal Group)

behind tubular transducers to widen the effective beam width of the transmitter and receiver transducers. A long-standing successor to the pioneering Deccalarm is produced by the US company Aritech. To reduce false alarm risks the system is designed to distinguish between an intruder and potential false alarm sources such as vibrating windows, the to and fro movement of lighting fittings and curtains in a draught. It does this by assessing a net change in target range before triggering an alarm. Draughts themselves of course, are largely ignored because of the use of the radar principle.

Discussion points

Like radio, ultrasonics is one of the hardest ideas to grasp, because the thing that makes it work cannot be 'seen'.

Discussion doesn't help too much until you have your own practical experience of what happens. There is no substitute for getting, say, a Doppler radar ultrasonic sensor with a battery and alarm indicator in a room on your own, or with a colleague who knows how to stand completely still. Work through this chapter and check as many of the features as you can. Don't forget to find out what happens when the sensor is mounted on the ceiling or a roof-truss looking down. What shape of transducer would be best for this arrangement? Would masking be a problem? Why would ceiling mounting have an even better chance of successful detection with freedom from false alarms than in the early days of the Deccalarm?

16 Microwave radar sensors

As we saw in Chapter 15, the operational need to find something better than standing wave detection led to the adoption of radar techniques, using ultrasonics because there was nothing else economically available. However, when the British Government scientist John Gunn discovered that he could generate small amounts of microwave energy at very low voltages with a tiny semiconductor diode instead of needing, as before, a high-voltage heavy thermionic klystron valve, one of the applications seen for the new Gunn diode was in intruder detection.

Successor or alternative to ultrasonics?

The timing of the invention of the Gunn diode meant that development of intruder detection using microwaves came after the private-enterprise development of the ultrasonic radar sensor described in Chapter 15. But the Government work was done in a radar laboratory, and it was probably inevitable that the microwave version also used the radar principle, although, as we shall see, it need not have done.

Both the ultrasonic and the microwave versions were developed in the 1960s, but before we can consider whether the microwave method is a successor to the ultrasonic, we have to be clear on what microwave sensors were designed to do. Quite simply but quite independently, they were designed to detect the movement of an intruder in a building, particularly in instances where the intruder had evaded any perimeter detection system, such as penetrating a door as illustrated in *Figure 15.1*.

So they were aimed at doing the same job as ultrasonic sensors, and although the design objectives were much the same in both cases, the microwave method, too, has been found in practice to have its strong and weak points. The difference in cost of microwave sensors and ultrasonic sensors as such is insufficient to dominate a decision one way or the other. As microwaves do not suffer attenuation in air, the usable detection coverage per microwave unit is larger than that possible with ultrasonics. But microwave sensors using the 2.4 and 10 GHz frequencies cannot be used safely near windows due to the false alarm risk of detecting something moving outside the window. With ultrasonics, there is no such risk.

Properties of microwaves

Already in this chapter some words and concepts have been mentioned which may seem unfamiliar. There is no reason why you shouldn't feel at home with this chapter, if you would please read or reread Chapter 4 first. To avoid too much repetition of radar ideas, it would help if you would read about ultrasonics as well, in Chapter 15.

Given that background, we can concentrate more on the differences between microwaves and other devices and on their respective pros and cons.

Energy

We are used to the idea that we can get heat from the Sun. The energy is radiated by the Sun and travels through empty space until it reaches our bodies and warms us up. Light is just the same type of energy coming from the Sun but it has a different effect upon the body, because it has a higher frequency and a shorter wavelength.

We can also produce this energy artificially. Just as we have artificial light, so we have other devices to produce heat. Microwaves are yet another form of the same type of energy, all being electromagnetic energy. The meaning of that word is unimportant to us, except to distinguish it from other forms, such as acoustic energy.

Wavelength

In spite of their name, microwaves are distinguished by having wavelengths longer than those of either light or heat, and can penetrate much deeper into human bodies and other materials than either. If that were not true, there would be no excuse for having microwave ovens – they cook quickly because the heat is generated immediately deeply in the food itself, and we do not have to wait while 'ordinary' higher-frequency heat is conducted slowly from the surface into the food. Wavelengths of 25 cm are typical in cooking, and microwave detection uses both this and also shorter and longer wavelengths, as explained later.

A word on safety

From time to time there are 'scares' about the possible hazards to human beings arising from radiation from microwave ovens, and the problem may be linked in some people's minds with intruder detection as well. In microwave ovens the powers involved are of the order of kilowatts, and in microwave sensors most Governments limit the usable power to not more than between 1 and 10 mW. So, the powers used for intruder detection are roughly one-millionth of those used for microwave ovens.

People must judge for themselves from the absence of reported incidents or from any authenticated reports if they do exist. With 20 years' and more of experience by designers and users to go by, I would say as a personal view that I am perfectly happy with the safety of microwave intruder detection at all wavelengths down to and including 3 cm (X band). But for shorter wavelengths, having worked on millimetric-wavelength radar for some years, I would resist the use of wavelengths of 1 cm and less for indoor intruder detection, on the grounds of prudence.

Velocity

Returning to other properties of electromagnetic energy: whether it is natural or man-made, the speed at which it travels from the source to the receiver is the same, the speed of light. Even now this speed of 300 000 000 metres per second seems unbelievable, especially when we think of the effort needed to push an aircraft beyond the speed of sound, at 332 metres per second. If only the air wasn't there.

The effect of air

The joy is that, so far as microwave energy is concerned, the air isn't there. Microwaves penetrate air as if it didn't exist, and at the speed of light. Thus, the snag that made the standing wave method of detection with ultrasonics so troublesome doesn't exist. So, for microwaves, the radar arrangement of transmitter and receiver, although attractive, is not necessary and the use of standing waves is entirely practicable.

Optical properties

If you know the properties of light, you know most of what you can do with microwaves. Microwaves travel in straight lines, so you need a line of sight between transmitter and receiver; but having said that, microwaves can be reflected, refracted and focused.

Penetration

We touched on the idea of penetration when we were looking at different forms of electromagnetic energy. But there is more to be understood about it in order to contrast and compare, choose and apply microwaves and ultrasonics.

The key fact is that microwaves will penetrate anything other than metal. Whether it matters or not depends upon the density and thickness of the non-metal. A brick wall, for instance, absorbs most of the energy from an intruder detector, so that false alarms due to legitimate movement on the other side of a wall are unlikely, especially if use is made of the optical properties and the sensor is placed not to look directly at the wall. Microwaves 'see through' most wooden doors, hardboard panels and glass, so the use of a microwave sensor close to a shop window would almost certainly lead to trouble.

Even ultrasonic energy can go through very thin materials such as single sheets of paper and plastics. But it cannot penetrate anything else.

As a way of remembering penetration, therefore, we can say that microwave energy can go through non-metallic solids 'because of its high velocity' but not through metal 'armour'. Ultrasonic energy, on the other hand, ambles along at a much slower speed and penetrates hardly anything.

Principles of operation

Would you accept that the creature which knows more than most about positional identification, obstacle and person location is the bat? In radar work I have turned to the bat and his ways again and again for inspiration, and the fact that he uses ultrasonics is encouraging but incidental, because much of what he does can as readily be done by microwaves. Up to a point, that is. Some of the things he is able to do are so advanced technically that humans tend to look for simpler ways that can be implemented inexpensively and to put up with the resulting limitations.

More about the Doppler effect

If it is worth his while, a bat can quite readily fly in the dark through a gap no wider than his wing span. To do that he must be able with his complex radar system to measure how much to the left or right he must steer, how far he is away from the gap and how wide it is, and how fast he is going, so that he can judge when he will arrive at the gap. He uses the Doppler effect to judge his speed, and depending upon the species of bat, he uses pulsed or frequency-modulated signals or a combination of both methods to measure distance and direction.

Fortunately, in intruder detection we do not need to know speed, nor really do we need to know the direction taken by an intruder, so long as we know that he is there and moving about in an area that can be identified in other ways. And Doppler signals are all we need to tell us that he is moving.

It is worth noting that in the case of the bat he is moving and the gap he is aiming at is stationary. With intruder detection the target or intruder is moving and the detector is stationary. The Doppler principle works just as well either way, as it detects only relative motion.

Radar detection

From Chapters 4 and 15 you will know that in radar the transmitter and receiver are mounted side by side, and that a signal is sent out continuously in a controlled direction by the transmitter. Anything in the path of the transmitted signal will cause some of the transmitted energy to be reflected back to the receiver. If the object in the path is stationary, then the frequency of the reflected energy will be the same as that transmitted, and the receiver will ignore it. In microwaves this is true even in vigorous air movement, so air movement itself does not cause false alarms with microwaves as it can with ultrasonics, where the radar cancellation effect described in Chapter 15 only helps for the smaller air movements.

If the object moves, as with, say, an intruder coming through a door, then the frequency of the reflected energy will be different from that of the transmitted energy, and the receiver will use that information to trigger an alarm signal.

Standing wave detection

Given that air movement in itself does not matter with microwaves, you might say, quite rightly, that it should be feasible to use microwaves in a standing wave arrangement. That there are so few systems using this principle probably stems from the problems encountered with the early UHF radio alarms, which suffered from the same trouble with dead spots as the ultrasonic standing wave systems covered in Chapter 15. If only someone would add the one or two additional receivers to give the diversity reception needed to overcome the problem, a very effective method of protecting warehouse risks would become available again.

As the problem of dead spots, or situations that can cause false alarms because of loss of received signal, can affect some forms of microwave fence in outdoor applications, the subject is referred to again in Chapter 19.

Evasion

To microwaves an intruder is no more than a bag of water, and water is an effective reflector of microwave energy, particularly if the water is not pure. So, in spite of the potentially deeper penetration by a minute amount of microwave energy into the body, there is no way that microwaves can see through and ignore an intruder.

Reliability and false alarm control

Much of what was said in Chapter 15 on designing to avoid nuisance alarms applies equally to microwave sensors. In particular, the signal-processing electronics can be almost identical.

A problem with microwave sensors working in the popular X band, at a wavelength of about 3 cm, is that the typical Doppler frequencies obtained fall within the mains ripple frequency bands of 50–60 Hz and 100–120 Hz. The exceptionally high mains ripple rejection necessary if it is not to interfere with very low level Doppler signals of similar frequency can cause design, and particularly long-term reliability, problems. Also, the Gunn diode, if used for microwave generation, is unfortunately inefficient, and deterioration in the thermal contact between the diode and the heavier metalwork of the resonant cavity can lead to overheating and failure. One way of overcoming the efficiency problem is to use one of the more recently developed sources of microwave energy, such as gallium arsenide field effect transistors (GaAsFET), to replace the Gunn diode.

Both the ripple frequency problem and the efficiency problem can be overcome by deserting 3 cm wavelengths in favour of 12 cm, thus quartering the Doppler frequency and taking it well away from the mains ripple frequencies. In addition, 12 cm can be generated very efficiently by transistors hard-soldered into the microwave circuitry, overcoming the risk of overheating. The merits of 12 cm are discussed further later in this chapter.

Going one better

A perhaps daring attack on the problems of using microwaves for intruder detection has been made by the UK company Guardall. Their models MX950 (short-range) and MX960 long-range) appear in *Figure 16.1*.

By using the exceptionally high transmitter frequency of 24 GHz (1.25 cm wavelength) the microwave penetration of glass is much reduced, with a corresponding reduction of false alarm risk. It can be seen that with careful selection and positioning of the detector, as far from any glass as

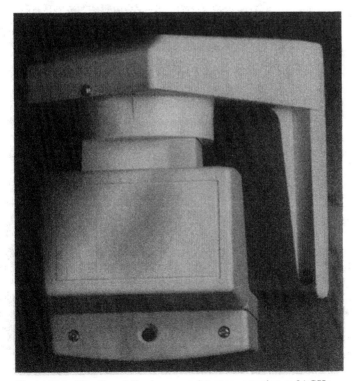

Figure 16.1   The Guardall microwave detector operating at 24 GHz

possible, the MX950/960 can enter the application high-ground uniquely held by ultrasonics for freedom from false alarms in glazed areas.

Beam shaping

Cost is such a factor in security that designers tend to use components that have already been developed for other applications. The disc frequently used as an ultrasonic transmitter and receiver in security and illustrated in *Figure 15.2* was developed for remote control of television receivers, and it just so happened that the 60° or so conical beam width of these discs gave very effective coverage and control against false alarms when used for intruder detection.

Likewise, with microwaves the most commonly available wavelength from other applications was 3 cm. Instead of using wires to connect one microwave component to another it was usual to 'pipe' the energy through a rectangular waveguide or tube, as illustrated in *Figure 16.2*. Waveguide tube of this kind was manufactured in large quantities and when it was realised that transmitting 3 cm energy through the open end of tube measuring about 2.5 × 1.25 cm gave a beam width of about 60 × 120°,

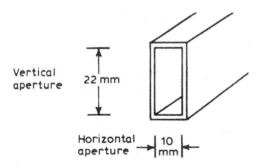

Vertical aperture   22 mm

Horizontal aperture   10 mm

Figure 16.2   Waveguide aperture used as a microwave 'aerial'. A single waveguide aperture 'aerial' may be used for both transmission and reception of microwave energy. Alternatively and preferably, for false alarm reduction, one waveguide aperture ('aerial') may be used for transmission (T), and a separate one for reception (R), separated as far as practicable in the vertical plane from the transmission aperture. (See also *Figure 16.6(b)*)

this arrangement was adopted, and no 'aerial' or beam shaping attachment was used.

You may be asking whether I have written that down correctly – does 60° correspond to 2.5 cm and does 1.25 cm correspond to 120°, or should it be the other way round?

Probably the simplest way of visualising the answer is as ripples on the surface of a tank of water. This method was used by Thomas Young to explain the behaviour of light waves in about 1801. Looking at the surface, as in *Figure 16.3(a)*, we can place a barrier across the tank with a small gap in it. If then with a straight paddle we create a wave motion in the water, we will see the waves travel uniformly towards the gap, but as the waves go through the narrow gap or aperture, they spread out quickly at

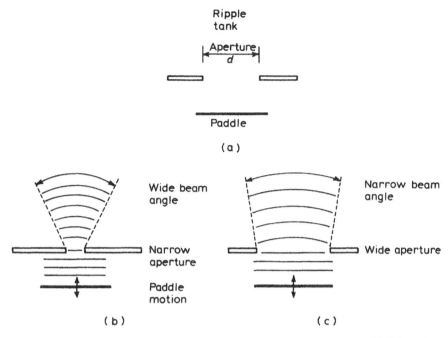

Figure 16.3 (a) Water tank for Young's demonstration of a beam shaping. (b) Effect of a narrow aperture *d* on the angle of the beam. (c) Effect of a wide aperture *d* on the angle of the beam. Note that the aperture size is added to the dispersion angle to arrive at the beam width at any given distance from the aperture

a wide angle, as indicated in *Figure 16.3(b)*. In *Figure 16.3(c)*, however, we have a wide gap, and the same wave motion will go straight through the gap, and through it will spread out only slightly. The bigger the gap the narrower the angle of spread. So, the above example of beam angles from a waveguide makes sense, even if it does seem perverse.

If you are getting the hang of the importance of wavelength for ultra-sonics and microwaves, you can think of it as the wider the aperture in terms of number of wavelengths the narrower the angle of radiation or beam width.

At microwaves, then, if we have an aperture which gives us too wide a beam width, all we have to do is to add a beam-shaping device called a horn, such as that illustrated in *Figure 16.4*. There is no need to go into the detail design of horns, but it is helpful to know two things about them.

(1) Beam width is inversely proportional to aperture. For instance, to decrease the beam width from, say, 80° to 20° we would increase the aperture by a factor of 4.
(2) Beam width is proportional to wavelength. This means that if we know what beam width to expect for a given aperture at 9 cm wavelength, the beam width would be reduced to one-third if the wavelength were reduced to 3 cm.

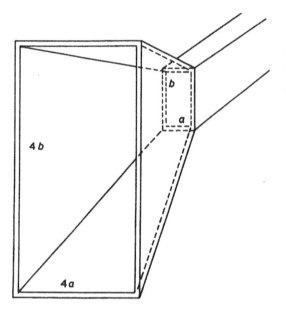

Figure 16.4 Using a horn for microwave beam shaping. The waveguide aperture has been increased by a factor of 4 in each plane, which gives a reduction by a factor of 4 in horizontal and vertical beam widths.

Coverage patterns

If we want to know whether a radar sensor mounted in a certain position would detect an intruder in all positions in the area where he is likely to be able to go, we are really asking, 'What is the coverage pattern of the radar?'

Although actual coverage patterns are three-dimensional, they usually have to be illustrated on paper, in two dimensions. This leads to two drawings, one of the horizontal pattern as seen looking down, and one of the vertical pattern as seen looking across. The patterns are usually pear- or apple-shaped, with the 'stalk' at the radar and the limit of detection distance at the 'flower' end, as illustrated for a typical indoor radar in *Figure 16.5*.

The sides of the pattern can usually be predicted from the beam width, but to get the overall shape there is no substitute for actual tests. It is usual to walk slowly to and fro, looking in the direction of the radar sensor and noting the positions when the sensor triggers. If evasion by very slow walking is a risk to be considered, the pattern tests can be repeated at the slowest speed envisaged, and also repeated by walking across the pattern, keeping the same distance from the sensor. This checks that the electronics sees the lowest required Doppler frequencies. If there are difficulties here, it may be helpful to use techniques that have had to be developed for passive infrared sensing. The vertical pattern can be found by mounting the radar on its side, so that, so far as the walk test is concerned, it is a horizontal pattern.

An interesting application of coverage patterns is discussed in the next section, on outdoor radar.

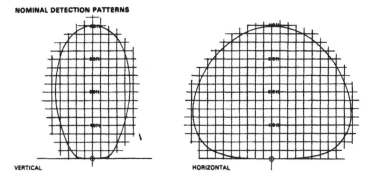

NOMINAL DETECTION PATTERNS

VERTICAL   HORIZONTAL

Figure 16.5   Typical detection coverage pattern of an indoor microwave Doppler radar sensor

Outdoor radar sensors

When we considered coverage patterns, it was emphasised that a human being was needed for the tests. This is because it is almost impossible to design an artificial man able to reflect the same energy as a real man – that is, having the same echoing power as a man. Nor can an artificial man generate the multitude of additional frequencies that come from limb movements and are important to a Doppler sensitive device. The smaller the man or woman the smaller the echoing power and the shorter the maximum distance at which they can be detected. But all can be detected close to the radar. Similarly, a bird, of yet smaller echoing power, can still echo enough energy when close to the radar to cause a false alarm. An understanding of coverage patterns was put to good use in an outdoor radar designed to overcome this problem.

Looking at *Figure 16.6(a)*, we see a typical vertical coverage pattern. However, the vertical pattern (*Figure 16.6(b)*) for outdoor radar is different from the usual. The key argument is that if a target such as a bird is not 'illuminated' by the transmitter, no microwave energy can be reflected from it to the receiver and it cannot cause a false alarm. The same applies, of course, if the bird is in the coverage pattern of the transmitter only. Any microwave energy echoed from the target cannot be picked up by the receiver. At distances further from the radar, where the vertical coverage patterns merge, reflections from birds can give a Doppler signal, but given that the intersection distance of the patterns is well chosen, the signal from the bird will be too weak to cause a false alarm.

Reflections from human beings are strong enough to cause a real alarm wherever the transmitter and receiver patterns overlap. Separation of the patterns is achieved by separating, one above the other, the transmitter and receiver parts of the radar sensor, by a distance very much greater than normal, a guide figure being 100 wavelengths or, say, 3000 mm (10 ft) for X band. For longer wavelengths compromise design starts to be necessary to make the system practicable, and this, in turn, is compensated by lower sensitivity to small targets.

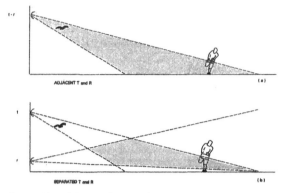

Figure 16.6 (a) A typical vertical coverage pattern for an indoor microwave radar sensor. (b) For an outdoor microwave radar, the pronounced separation of the transit and receive heads can reduce false alarms

When X band radar is used out of doors, false alarms from rain and hail can be troublesome, but the above arrangement of well-separated transmitter and receiver is also helpful in rejecting rain problems.

Suitable and unsuitable applications

In discussing applications of microwave sensors it is necessary to assume that the sensors are properly sited for those applications, as indicated below.

Large areas

In general, microwave sensors are able to cover a greater area per sensor than any other method of area or volumetric detection. For still greater area, the Government-authorised frequency bands are wide enough to enable several sensors working together in the same large area to be selected to work at individual frequencies within the band. This avoids the risk of two sensors inadvertently operating on almost identical frequencies and so causing an apparent Doppler signal and consequent false alarm.

The benefits of microwaves are improved further in large areas by using ceiling or roof-truss mounting, which gives coverage per unit up to twice that obtained by wall or column mounting. Given the normal beam shape of 120–150° in one direction and 60–75° in the other direction, no energy need be directed at walls or windows or even doors, which all pose false alarm risks.

When the longer wavelength of microwave sensors compared with ultra-sonics is taken into account, the sensitivity to vibration is less, and, in general, microwaves suffer least from environmental problems. And because microwaves can 'see through' thin materials such as draught-disturbed paper and cardboard, and because their 'transparency' increases with wavelength, the longer the wavelength within reason the better for avoiding typical warehouse false alarm problems.

Medium-sized areas

A medium-sized area is rather harder to define than large areas requiring multiple sensors to give adequate coverage. A medium-sized area is probably best thought of as needing only one sensor for full coverage, with possibly a second to fill in a blind spot. The choice between the various forms of space detection is a personal one, although the characteristics of the area may rule out one or more methods, depending upon the risk of false alarms.

In office areas, for instance, there are so many things that could cause an ultrasonic sensor to give false alarms that a microwave sensor mounted to avoid looking through glass and passage partitions may be the most suitable.

Small areas

As the area becomes smaller and the risk becomes higher, the construction of the walls, ceiling, doors, windows and floor all grow in importance, because they are closer to the sensor and become greater potential sources of false alarm. If the area is none too strong, with windows as in, perhaps, a shop, then ultrasonics sited to look at as little of the area boundaries as practicable would be preferable to microwaves, but the stiff structures, particularly those without glass, are better with microwave sensors. If the risk being covered is not too high, passive infrared detection, described in Chapter 17, would be an acceptable and cheaper alternative to microwaves.

Although other methods of detection can be cheaper than microwave, microwave detection is holding its own in particular for the higher grades of risk, and in that context it is probably wiser not to say too much.

Reference will be made in Chapter 17, on passive infrared detection, to the use of the so-called dual technology. In this technique a combined detector is made using passive infrared, together with ultrasonic, microwave or microphonic detection. Both detection methods have to trigger for an alarm to be raised. See Chapter 19.

Discussion points

Mention has been made of the influence of fashion in the selection of intruder detectors. This applies particularly to space detection, with the choice swinging between the three main contenders – ultrasonics, microwaves and passive infrared. For rational decisions to be made it is important to keep abreast of experience in the field with each of these, and to keep a look out for new contenders. Experience has to be matched against unit cost, total cost, certainty of detection, false alarm risks and the much less controllable preferences of the customer. With so many variables the only practicable way, it seems, of finding the best combination at any given time is to have organised discussion groups from time to time to pool their knowledge, thinking and experience. These discussions should give management a firmer base to operate from, remembering the useful concept that 'Standardisation is a guide for the time being'.

17 Passive infrared sensors

In Chapter 14 the use of an infrared transmitter together with a receiver to form an active infrared beam detection system was discussed. The word 'active' is included in the name to help to distinguish systems using transmitters from an alternative method of using infrared, without transmitters, called passive infrared.

You know that many of the most useful inventions were discovered by accident. It is said that an alert engineer, examining an active beam system after a bulb failure, thought that he detected a minute signal at the receiver when a colleague walked across its face. Whether this is true or not, or whether independent knowledge that human beings themselves are very effective transmitters of infrared energy (in the form of heat) led to the invention of passive infrared intruder detectors, no one seems to know for sure. Probably two or more quite separate people working independently in different ways hit upon the idea at about the same time, and probably there is some truth in both stories. Instinctively one feels that this has to be an ideal method of intruder detection.

The opportunity and the challenge

If the environment into which a villain intrudes is at the same temperature as his or her body then passive infrared detection is a non-starter. The opportunity comes from the fact that the environmental temperature is so rarely equal to the body temperature. The challenge comes in finding an effective method of sensing or measuring a temperature difference. I first became aware of its significance to security when I read of an inventor in the USA who claimed to be able to detect a dog wagging his tail from a distance of 100 yards. 'Heaven forbid,' I thought at the time; it looked like a guaranteed false alarm generator. Nevertheless, I wrote to him but got nowhere, as it was evident that the components to make a practicable system were not then available. In time, these did become available from development work particularly in Germany, the USA and the UK.

Sensing elements

Although it is results that matter to the user, it may be of interest to note that development in different countries led to preferences for different materials for the sensing elements. With its long history in infrared research, Germany concentrated on using lithium tantalate for its sensors, while the UK, probably with Holland, was more at home with lead zirconium titanate ceramics.

If it hasn't happened already, experience suggests that the industry will converge on one or two preferred materials, hidden, perhaps, behind non-meaningful trade names. Japan, for instance, seems to be thriving on something they call ferro-electric ceramic.

In order to achieve adequate detection sensitivity the heat energy 'seen' by the sensor needs to be fluctuating rather than steady. Fortunately, in the application to security, a moving intruder provides the basis for meeting that requirement. Also, the energy needs to be focused to some degree on to the sensing element. However, infrared energy does not pass readily through glass, so ordinary lenses cannot be used for focusing, and other methods have had to be developed.

Focusing elements

Two methods are used to overcome the limitations of conventional glass lenses. First, let us consider mirrors. Readers will remember the fun to be had in amusement parks with distorting, or anamorphic, mirrors. If we assume that an intruder is tall and thin, then one distorting mirror can make him or her look equally short and plump. If the short-and-plump image is now reflected onto another distorting mirror, placed at right angles to the first, the image can be shrunk to a dot; that is, it has been focused. *Figure 17.1* gives some idea of how this works.

Effective though it is, this method structurally is insufficiently flexible to meet more than a few of the varied protected area coverage patterns required. This need for application flexibility led to the adoption of the Fresnel lens technique, seen in lighthouses and searchlights, not to mention the focusing screens of some single-lens reflex (SLR) cameras. A section through a small such lens for use in PIR sensors is shown in

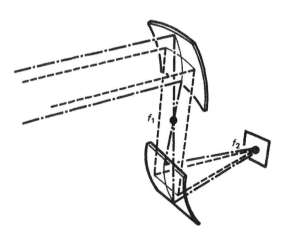

Figure 17.1 Precision optics and anamorphosis. (Supplied by Cerberus Ltd)

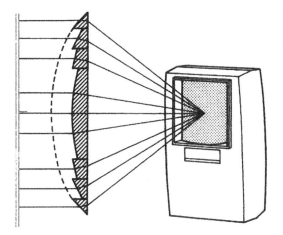

Figure 17.2 The Fresnel lens optical system. (Supplied by Cerberus Ltd)

Figure 17.2. Whatever type of lens is used there will be some loss of energy through it, and therefore some loss of detection sensitivity. The dotted line shows how thick a conventional convex lens would be. By using the stepped, Fresnel lens technique the effective thickness, and therefore loss, can be reduced significantly. Also, by using plastic instead of glass adequate efficiency and focusing can be achieved. As an example of design flexibility individual lenses for, say, 24 separate detection beams can be moulded into a plastic lens of size approximately 50 × 40 mm (Chartland Electronics Ltd).

One-, two- and four-element sensors

The early passive infrared sensors tended to use only one pyroelectric sensing element per intruder detector assembly. Applying the dual question 'Does it work – can it be made not to work?' soon revealed that, however good it was as a detector, it was prone to false alarm. The sensor manufacturers provided an answer to the problem by producing sensors incorporating two, or dual, elements. If one element is designed to produce a positive voltage on receiving heat and the other a negative voltage, then if the elements are connected in series (or parallel) and both receive heat simultaneously the voltages cancel out and no alarm signal is generated. The design of the sensor/lens combination has to be such that the heat from an intruder must affect only one sensor element at a time to generate an alarm voltage. On the other hand, changes in ambient temperature, acoustic noise and sunlight should affect both sensor elements simultaneously, and be cancelled out.

Figure 17.3 illustrates the terminology relating to one-, two- and four-element sensors. When comparing the detection to false alarm performance of the different configurations it is important to remember that PIR sensors respond best to movement across the 'fingers', while ultrasonic and microwave Doppler radar systems are most sensitive to movement towards or away from the sensor.

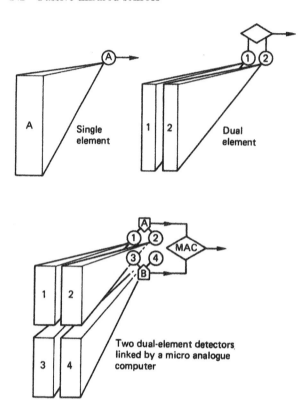

Figure 17.3 The terminology relating to one-, two- and four-element sensors. (Supplied by Pulnix)

Good though the dual element is, some obstinate false alarms remain. This has led to the use of four sensing elements. For instance, the Pulnix version, called the Quad Element Detector, uses two dual pyroelectric elements. These two outputs are fed to a signal-processing unit which triggers an alarm only when two signals from the quad system exceed a predetermined threshold level.

Masking

Passive infrared sensors tend to be resisted by system designers for higher levels of security because it is said that they are too prone to malicious masking. Some efforts are made in sensor housing design to limit the opportunities for hanging masking material over them, and there is much to be said for mounting sensors on ceilings (*Figure 17.8*). However, even then they can be sprayed with masking material – given that the villain can get into position to do it without being detected.

If that makes you sceptical about using PIR detection, I would urge you not to leave it at that; PIR is a fundamentally sound method of detection. If today it has difficulty in satisfying the dual question, 'Will it work; can it be made not to work?' there is a great deal of development work being devoted to satisfying users. One supplier said in 1997 that their product 'achieves the highest level of anti-masking protection available today.' That seems an honest comment, and implies that in 1998 and beyond things will be better still.

When considering the value of development work, remember Watson-Watt's famous saying, 'The best is the enemy of the good.' He had to rein-in his enthusiastic team of radar pioneers, otherwise he would never have had a working radar in time to help win the 1939–45 war. He did get something out and working, in time, but development did continue, at a terrific pace, and the demands for yet more advanced radar were met too.

Applications of PIR detection

Suitable applications become apparent as one understands the nature of the various coverage patterns possible with custom-made Fresnel lenses. The basic pattern is seen if you hold your hand horizontal and spread the fingers. *Figure 17.4(a)* shows a typical pattern for 90° horizontal coverage, though the angle can be anything from a few degrees to 180° with wall mounting and to 360° with ceiling mounting. *Figure 17.4(b)* shows the

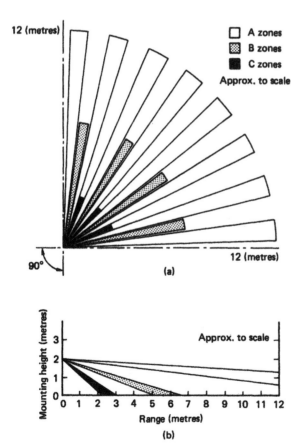

Figure 17.4 (a) and (b)

same fingers, tilted vertically at various angles to maintain floor coverage. Some models look vertically down as well, to minimize the risk of someone crawling close to the wall on which the sensor is mounted. These applications tend to be much the same as those for ultrasonic and microwave Doppler radar sensors.

Perhaps of more distinctive interest as an application of PIR sensors is the so-called curtain or 'thin-slice' coverage. The dual sensing element sees only one finger, made up of two closely spaced sensitive areas, as indicated in the plan view of *Figure 17.5(a)* Although thin, the electronics can be arranged to sense entering and leaving each zone. Ninety degrees of coverage is achieved in the other plane, as shown in the side view of *Figure 17.5(b)*, where the risk protected might be an art gallery, a staff entrance, or an office or shop window. By turning the curtain sensor to give a horizontal thin slice such vulnerable features as skylights in buildings can be covered. Yet another use, of value particularly in domestic applications, is where the sensor has been turned upside-down, to give a curtain falling short of the ground or floor. With fair judgment of the ground clearance needed, allowing for animals jumping onto furniture, the domestic risk of false alarms from animals can be reduced. The detection coverage of windows and doors can be maintained, so the void near floor level can be acceptable in most domestic risks.

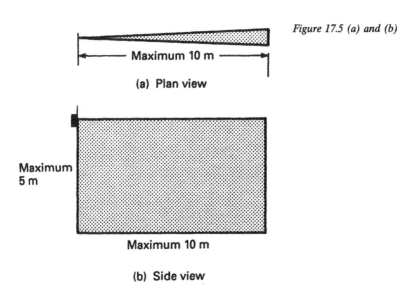

Figure 17.5 (a) and (b)

Maximum 10 m

(a) Plan view

Maximum 5 m

Maximum 10 m

(b) Side view

It always helps if a security device can do something else useful as well. The most widely used application of passive infrared sensors may be out of doors, when coupled with lighting switched on as the sensor triggers to the approach of friend or foe. To the friend it is a welcome: to the foe it is a deterrent.

Figure 17.6 *'Sentor 640' portable PIR detector, usable indoors or outdoors, singly or in multiples, with radio alarm signalling*

Figure 17.7 *Wall or ceiling mounted, the Reflex PIR detector offers good styling and optimised performance in over 2 million installations in five countries. (Supplied by Texecom)*

Figure 17.8   *A ceiling-mounted passive infrared sensor. (Supplied by Aritech)*

Figure 17.9   *A passive infrared sensor integrated with floodlights and a TV camera. (Supplied by GJD Ltd)*

What of false alarms?

PIR sensors benefit from the experience gained in dealing with false alarm problems in ultrasonic and microwave Doppler radar sensors. Such things as radiated and mains-borne interference, vibration, central heating radiators and excess detection sensitivity are all common to PIR sensors as well, and methods of treatment are established.

Less common problems that have proved tiresome in PIR-based systems include bright sunlight and car headlamps, audio and subaudio sound. Sunlight containing sufficient infrared can get through glass windows, while PIR sensors act also as piezoelectric audio receivers. PIR sensor circuitry has to be responsive to frequencies from virtually zero upwards, a range cut off as unnecessary in other types of sensor. Two- and four-element PIR technology is doing much to deal with the relatively unique problems of PIR systems. For you to decide whether enough has been done to defeat PIR false alarms, as with four-element technology, or whether you would rather hedge your bets by using dual technology as described in Chapter 19, it could be helpful to read the factors given in 'Discussion points' in that chapter.

Discussion points

If fashion is a factor in the decision process for intruder detection device selection, where does PIR stand? If it is winning the gold, is it on grounds of lower unit and installation cost, or has it operational advantages over other devices? Or has fashion nothing to do with it now? Isn't it more definitely a question of what the sensor is required to do – to protect the average risk or a high risk?

18 Microphonic sensors

The prefix micro- (from the Greek *mikros*, small) is rather overworked in the electronics business, and the fact that it means small does not always help in an understanding of the function of the device it is linked to. The word 'microphone', however, is in such common use in association with radio, television, telephones and public address equipment that it is absorbed without question into the language. For our purposes, a microphone is a type of transducer for converting small amounts of acoustic and mechanical energy into electrical energy which can then be amplified and used elsewhere.

Geophones

Microphones were not developed for security, nor were geophones. These came into being as special types of microphone for exploring the depths of the Earth for oil and other minerals. They have to be both very robust to withstand the initial shock as an explosive charge is set off at the Earth's surface, and very sensitive to detect the minute echoes reflected radar-wise from the various strata down below.

They enjoyed initial popularity when first applied to security for perimeter fences. On fences they had the advantage in principle of being sensitive in the up and down directions, but not side to side, which suggests a capability for discriminating against wind effects. With experience it became evident that they were too sensitive for security, which led to their being grossly overloaded by most wanted and unwanted signals, which masked the otherwise valuable properties of discrimination.

The most commonly used of the early types were made by the French firm Sercel, and, later, versions became available in which the sensitivity had been modified to match the security environment. Incidentally, this situation illustrates something that happens quite often in security. Sensitivity tends by accident or design to be much too high for the environment, and certainty of detection can be achieved with much reduced risk of trouble if the sensitivity is reduced.

An interesting attempt was made in the Sercel equipment to give an indication in the local security office of the intensity of detected signals by providing an illuminated 'thermometer' gauge for each zone. This gave visual warning of the location of a possible attack before the signal became great enough to raise a full alert.

Piezoelectric sensors

It is a fortunate fact of nature that when various minerals such as quartz are squeezed they can generate electricity. One or two mental pictures may be helpful here. Imagine any small object such as a wooden or plastic dice resting on a polished table top. If we press the side of the dice it will move with no apparent resistance to the pressure applied. If we press down on the top of the dice, it doesn't move at all, but we can imagine that the dice might be squashed a little. Now, instead of resting on the table top, let us imagine that it is glued to the inside surface of a shop window, almost anywhere on it. Then we bang hard on the outside of the window. Hopefully, the glass doesn't break, but it will flex inwards. The effect on the dice is negligible, much the same as when we pushed it across the table top. To squash the dice, it would be necessary to have something solid on the other side of it to 'resist' the movement of the glass. If a piece of fairly heavy metal is glued on so that the dice is sandwiched between the glass and the metal, then we can see that the dice would be squeezed if the window was banged, due to the inertia of the metal.

The significance of all this is that if the dice on the window is replaced by a piece of specially prepared quartz a voltage is generated between the faces of the quartz being squeezed, and this voltage can be used for intrusion detection.

One advantage is that the voltage is only there when the squeeze is increasing or decreasing; another is that the voltage increases with the rate of change of squeeze. In other words, the voltage depends upon acceleration, which in physics is often abbreviated to the letter G, from which proprietary names such as G-Tector are derived.

In practice, a synthetic material such as barium titanate is used instead of quartz as the piezoelectric sensor, the same material as is often used in ultrasonic detection devices, described in Chapter 15.

Breaking-glass detectors

We talked above about banging the shop window. It may be useful to know that this was happening, but more likely it would be logged as a false alarm. Much more important is to know whether the glass has been broken. This we do by making use of the G-sensing capability of piezoelectric sensors. When glass breaks, it generates a complex spectrum of high audio and low ultrasonic frequencies due to the massive accelerating and decelerating motions involved at the instant of breaking. Piezoelectric devices generate their maximum voltages under these conditions, and are thus able to discriminate quite easily between real and false alarm situations.

Fence and wall sensors

Owing to the very high discrimination between low frequencies and high frequencies achieved by acceleration-sensitive materials such as piezo-electric crystals, they are potentially useful for use out of doors on fences and walls. Most legitimate vibrations are of relatively low frequency, while most intrusion and attack frequencies are high.

As the voltage generated is proportional to acceleration, G, we can look at it the other way and find what movement is needed at any given frequency to generate a given voltage. This gives us a measure of the discrimination available.

To generate a voltage V, we need a movement of, say, one-tenth of an inch at 10 Hz, and this leads to one-hundredth of an inch at 100 Hz and only one-thousandth of an inch at 1000 Hz.

The preference given by piezoelectric sensors to sharp signals from, say, cutting a wire fence is quite impressive. However, few firms have availed themselves of these properties, even though they give all the advantages of the inertia switch without the attendant contact problems.

Electret cable sensors

'Having one's ear to the ground' means keeping well informed, and that is what we need to be in security. The electret cable is a micro-phone, specially designed very much with security in mind (US Patent No. 3 673 482). Like the geophone and like piezoelectric sensors, the electret cable needs to be in contact with whatever can give it infor-mation, whether that be a fence, the ground or roadway.

Where it differs from the geophone and piezoelectric sensors is that it reports faithfully all it hears. You will remember from earlier in this chapter that piezoelectric devices when used as accelerometers or G-sensing devices tend to ignore lower frequencies in favour of higher frequencies, just as inertia switches do, and geophones are much more concerned with sensing low frequencies.

We have seen that although it can be helpful if the sensor itself helps to filter out false alarms, it is of no help at all if it also filters out infor-mation we need. The electret cable tends to report all, leaving discrimi-nation entirely in the ingenious hands of the electronic circuit designer.

If you know how the electret microphone as used in public address and broadcasting works, you may be thinking that this section should be under the heading of 'Field-effect sensors', to be discussed in Chapter 21. You would be right if this was a book on electronic techniques, but, in fact, it is about applied electronics; and where a device can be related to something we know in everyday life, it seems to help understanding if we think first of what a device does and only then go on to see how it works.

The electret is the dielectric material of a capacitor which has been given a permanent pre-charge by applying a high polarising voltage to the dielectric when it is heated to near melting point. The discovery that chemicals in the fluoroethylene group can be made to hold an electric charge in this way had made obsolete the old way of making condenser microphones with a continuously applied external high voltage. The fact that this special type of electric field behaves like a magnetic field is reflected in the name, since the name 'electret' is derived from the two words 'electric' and 'magnetic'.

A typical section of the cable is shown in *Figure 18.1*. Minute pressure from, say, a wire in a chain-link fence caused by someone climbing or cutting the fence is enough to generate a signal that is transmitted to the control position. Clearly, if a signal can be generated so simply, then a vast number of varied signals must be generated along a perimeter using an electret cable for intruder detection.

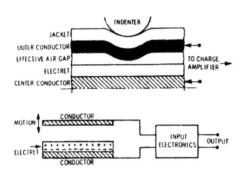

Figure 18.1 A section through an electret cable and diagram showing the principle of operation. GTE Sylvania patent. (Supplied by Fieldtech Heathrow Ltd)

The skills come in the design and testing of the signal-processing electronics, almost all of which is done before the equipment is supplied to a customer. It has been necessary from predictions, experience and many site trials to establish the range of false alarm creating situations, to record the electrical signature of these situations and to build electronic mirror images of them to aid in cancelling them out. Similarly, the signature of wanted signals, from intrusion attempts, are recorded, and ways found of enhancing them to achieve maximum discrimination between wanted and unwanted signals.

Although I have been able to carry out only laboratory tests on the GTE Sylvania Fence Protection system, illustrated in *Figure 18.2*, I was

Figure 18.2 An electret cable sensor installed on a chain-link fence. (Supplied by Fieldtech Heathrow Ltd)

particularly impressed when it rejected each attempt I made to cause a false alarm, and yet it detected reliably each attempt at intrusion.

However, remember the discussion we had in Chapter 6 regarding 'no man's land'. It is just asking for trouble from false alarms if any form of fence sensor is fitted to a fence on one side of which people can legitimately be. A fence within a fence is needed, even if the separation possible is only half a metre of less.

Conventional microphones

Partly in their own right and partly in the context of dual technology as discussed in Chapter 19, conventional microphones are enjoying a new lease of life in security applications.

Where only a limited number of areas (say, within a building) have to be monitored, microphones covering the audio range of frequencies have long been an invaluable intruder-detection device. When signals are amplified and heard in the site security office they need no interpretation to be intelligible.

It was, of course, impracticable to extend directly listening to multiple sites, each feeding a central monitoring station. The breakthrough came by:

(1) Treating the microphone like any other security intruder detector, with the audio sound rectified to give an on–off signal to the central station.
(2) Having got through to the central station, the security officer can then switch the actual audio signal causing the alarm through to a loudspeaker. He can then say without doubt whether the alarm is real; he can listen to whatever is going on in the protected premises.

With positive information like that to go on, the central station can brief the police accordingly, and an arrest is that more likely. Similarly, it is much less likely that the central station would pass on a call for the police that turned out to be a false alarm.

The market leaders in this method of using microphones seems to be the American firm Sonitrol, who have installations in many countries, ranging from strong-rooms to schools.

In the UK, Verified Systems have joined forces with Pyronix Ltd with similar objectives but different methods.

Discussion points

There was a period when microphone sensors tended to be neglected by equipment and systems designers. The continued false alarm problem has led some to rethink former techniques, and the prospect of using microphones both as initial detectors and for verification seems to offer an enduring future.

It is unusual for something to be designed specifically for intruder detection, and the electret cable is one of the attractive exceptions. If you have not met it in practice so far, endeavour to set up a trial installation and satisfy yourself whether it does or does not advance the art beyond your present knowledge. Compare your findings with other people's opinions, and, if favourable, move on from there to consider with them what other applications of the electret cable may be suitable.

The same procedure might be followed for microphonic sensors. Discuss with colleagues whether a potentially winning technique is being neglected.

19 Dual technology

The false alarm problem

To the police a call is a call, real or false, but there is nothing to tell them whether a call is real or false, so they have to treat all calls as real and find out for themselves.

As discussed elsewhere, so many of the calls prove to be false compared with the real calls that the resulting gross waste of police effort has led them to refuse to attend premises with excessive false alarms (see ACPO policy).

This refusal is unacceptable to insurers and occupiers of protected premises, so the burden of remedial action is thrust back on the designers, suppliers, installers and maintainers of the intruder detection systems. But no new detection system capable of meeting legitimate police demands emerged.

When all practical steps have been taken, there is one factor remaining – the environment, and that cannot be changed.

It has been shown in previous chapters that the various methods of intruder detection differ, in that one type may be prone to false alarm in one environment, while another may be immune. Given that both can detect the movement of an intruder, why not use one of each type of detector together? If both must trigger to give a real alarm a whole category of false alarm can be avoided.

Two of a kind?

Some say, 'why not use two identical sensors side by side?' That would be to solve the wrong problem. Duplication can be the right answer if it is equipment reliability that is the problem but with mean-time-between-component failures now being measured in years rather than hours, it is not clear how quick results on the scale required can be achieved by concentrating on the equipment itself. So we have to get our minds back to the environment. In this way an idea we call dual technology was born.

The one method that seems sufficiently different from the others is PIR. Doppler systems are at their best with motion towards or away from them. PIR prefers motion across its field of view, thus missing, or rejecting, a potential source of false alarm. Both detect humans because, besides bodily movement, there is also limb movement in various directions simultaneously.

Examples follow of how various firms have interpreted the concept of dual technology.

Proprietary equipment

The combinations of detection technology referred to below are:

(1) Ultrasonic Doppler with PIR.
(2) Microwave Doppler with PIR.
(3) Microphonic with PIR.

Ultrasonic with PIR

The sensor shown in *Figure 19.1* is from the Aritech Type DD 325/335 series PIR detector combined with an ultrasonic Doppler detector. Both systems must detect motion within a given time for the alarm to be triggered. Susceptibility of the two to false alarms tends to be different, so generally only one or the other detector triggers but no alarm is set off. The PIR sensor has its own anti-false alarm feature of a differential PIR element, and can be installed to give either a simple five-finger horizontal pattern with floor coverage or a vertical curtain pattern. The ultrasonic

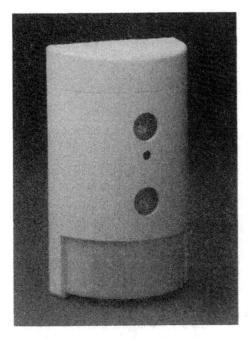

Figure 19.1

part of the system gives a volumetric Doppler detection pattern, and detection sensitivity is adjustable out to 5, 7 or 10 metres.

Microwave with PIR

Many security people say 'If only microwaves didn't penetrate glass, microwave plus PIR used in a dual system would make the ideal intruder detection system'. That dream has come much nearer to fruition since components for generating and receiving 24 GHz microwave energy became available.

The firm C and K systems have for a long time concentrated on dual systems to combat the false alarm problem, and their adoption of 24 GHz (K Band) has nothing to do with their name. Just as in colours, different colours are given names, so in microwave jargon, different bands of frequencies are given letters for identification such as S Band (10 cm wavelength), X Band (3 cm), K Band (1.25 cm) and Q Band (8 mm). The virtue of K Band is that its quarter wavelength is usefully near to the typical thickness of glass in windows and the microwave energy tends to be reflected rather than transmitted through the glass. The C and K implementation of this technique is shown in *Figure 19.2*.

Microphonic with PIR

The third category to have emerged in dual technology is the combination of microphonic detection and PIR. Of its several applications, its most

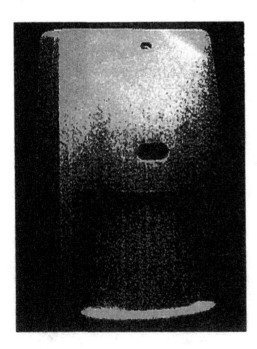

Figure 19.2 The C and K model DT-700 sensor using PIR together with extra high frequency microwave to limit wall and glass penetration.

likely strength is in lock-up-and-leave premises that are prone to entry via a broken window.

The first of the two sensors is an electret-type microphone with a bandpass filter tuned to accept the sounds of breaking glass, principally in the audio frequency range of 6–8 kHz. There are so many other sources of sound in this frequency band that, on its own, the breaking-glass detector could generate more false alarms than real. It is coupled therefore to its 'dual' partner, but in an unusual way. That partner is a PIR sensor, set to look generally at the window area. When the glass detector triggers it does not raise an alarm, but it does 'arm' the PIR sensor. Even if villains who break the glass wait a long time, 'until the coast is clear', as soon as they enter through the broken window the PIR is ready and waiting for them, and immediately raises the alarm.

Discussion points

It may be just worth reading again what I had to say earlier in this chapter about the distinction between certainty of detection and freedom from false alarms. It needs quite clear thinking to decide what to do, and you may want to check whether my reasoning is correct. What does stand out is the need first to decide which objective has your priority. If you are tempted to think that quad or two-channel dual signal processing is a better answer than dual technology you may well be right, but can you explain why? There is no doubt that dual technology is aimed at reducing false alarms, even if C and K Systems Inc. do use some of the safety factors gained by dual technology to increase their usable detection sensitivity.

If, however, your risk is so high that you lean towards making certainty of detection your priority, would you choose dual technology or effectively four-channel discrimination with one technology – namely, PIR?

20 Microwave fence sensors

The name 'microwave fence' gives a very good idea of what the sensor is intended to do, although the coverage pattern, if visible, and the electronic equipment that produces it bear little resemblance to what we think of as a fence. Compared with the majority of devices used in intruder detection, the microwave fence is one of the newer concepts.

The need for microwave fences

The microwave fence in its various forms was developed primarily for outdoor use, but there are situations where it can be useful also indoors.

Out of doors a common need is for a first line of defence to be set up at the outer perimeter of a site to give early warning of intrusion. The natural thing to do is to attach vibration sensors of one kind or another to the boundary fence itself. But try as one will, it does not seem possible to eliminate false alarms caused by people legitimately on the outside of the boundary fence. No doubt, determined souls will go on trying, but meanwhile others will be looking for alternatives.

Among alternatives, a system which became and remains a front runner consists of several beams of infrared light, stacked one above the other and mounted just inside the boundary fence, as mentioned in Chapter 14. As explained there, infrared beams can be interrupted by fog, and the need to avoid false alarms from birds often meant operating two beams in parallel. The reason for the trouble with the infrared light in fog is that its wavelength is so short that it is about the fog droplet size. The energy is absorbed before it reaches the receiver and the system goes into a continuous alarm condition until the fog clears. In many locations the likelihood of fog is so rare that this limitation does not matter. But in locations where it does matter something else is needed. The natural thing to do was to increase the wavelength, and because microwaves see right through even dense fog, the microwave fence concept became an alternative to perimeter fence-attached devices.

The ways in which the detection zone in microwave fences is controlled becomes more interesting the further one delves into it.

Evolution of three basic types of microwave fence

Not surprisingly microwave fences use the beam principle, and initially they were almost like infrared beams.

In addition to being perhaps the instinctive thing to do, when trying to improve something it is important to sort out whether the principle upon which it worked was wrong, or whether the principle was right but the application of the principle was wrong. In the previous paragraphs it was said that the basic thing wrong with the infrared beam was the wavelength. This is a fair way of saying that in some circumstances the application of the principle was wrong but that the principle itself, the use of beams, was right.

Beam-breaking system

The beam-breaking arrangement is illustrated in *Figure 20.1(a)*. The resemblance to infrared beams is clear. The transmitter and the receiver units are set up facing each other, at any spacing required up to the maximum recommended by the manufacturers. This might be 100 m or more technically, but operationally the spacing is more likely to be limited by the need for a clear line of sight between each transmitter and receiver. The line of sight can be interrupted by changes of direction of the boundary fence, by undulations of the ground and by the usual run of obstructions.

At each significant change of direction, up or down or sideways, a further transmitter pair is needed. The height of the units above ground is typically about 1 m, and if the system operates at a microwave length

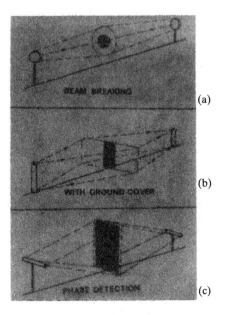

(a)

(b)

(c)

Figure 20.1 (a) A microwave fence using the beam-breaking principle; (b) a beam-breaking microwave fence with vertical aerials to give increased ground cover against crawling intruders; (c) an alternative principle is phase sensing, which uses horizontal aerials with wide vertical and narrow horizontal beam widths, to give extended ground and high-level cover

of 3 cm (X band), a circular bowl aerial of 25 cm diameter will give a conical beam width of nearly 10° – that is, 5° above and 5° below a line joining the centres of the transmit and receive aerials.

Looking again at *Figure 20.1(a)*, and bearing in mind that 5° downward coverage, the similarity to the infrared beam is emphasised. Both are beam-breaking devices for detection, and in neither case is coverage at ground level a primary objective.

Beam breaking with ground cover

If it is felt that the coverage pattern using conical beams described in the previous paragraphs leads to excessive risk of an intruder's evading the beam by crawling under it, then further beam shaping can be beneficial.

One approach is to use a vertical parabola-shaped cylindrical reflector about 1 m high and about 25 cm wide behind a long slotted waveguide-type aerial. If the base of the aerial is at ground level, as illustrated in *Figure 20.1(b)*, then the problem of evasion by crawling can be solved, given that the ground is reasonably level.

This arrangement can give a high probability of detection with few environmental problems to cause false alarms. Except, that is, from inadvertent detection due to the wide beam spread in the plan view – on either side of a line joining the transmitter and receiver.

By turning the aerial and reflector of *Figure 20.1(b)* through 90° and mounting the now horizontal reflector, say, 1 m above ground we have the most interesting of all the microwave fences.

Phase-sensing microwave fence

Figure 20.1(c) shows what that further configuration is. It incorporates many of the lessons we have discussed for both ultrasonic and microwave space detection, and to illustrate its features some of them are repeated here.

Beam shaping As mentioned above, a major problem in finding alternatives to vibration sensors on boundary fences and to infrared beams just inside boundary fences is the very limited ground space within the boundary that can practically and economically be allocated to intruder detection. However, proper use of beam shaping can go a long way towards resolving this dilemma.

A useful yardstick which helps in judging the effect of beam shaping is that an aerial aperture of 2 m used at a wavelength of 3 cm gives a beam width of about 1°. From this you can reason that an aerial 1 m wide would give a 2° beam width at the same wavelength. Now with that beam width in the horizontal plane we can get very close to the boundary fence without trouble. If we wanted the benefits of a longer wavelength, say 12 cm, then a 2 m aerial would give a 4° beam width, still quite usable, although it would need more of the valuable space to be sufficiently clear of the boundary fence.

When one looks at the horizontal transmitters and receivers as shown, it seems to be asking for trouble from intruders crawling under the beam. Two things are done to trap an unwary intruder. First, the aerials are made narrower in the now vertical plane, to give wider beam widths between, say, 15 and 20°. Wide beam widths touch the ground very shortly after leaving the aerial, and also reach higher upwards. Second, the standing wave principle is used to good effect.

Standing wave detection We discussed standing waves in Chapter 15 when finding how ultrasonic radar came into being. There we found that air movement can have a drastic effect upon ultrasonics. But with microwave detection air movement has no effect whatever, and most of the properties of standing waves are helpful.

To remind you of what is happening, I would ask you to think about the problem of fading in radio reception. Fading is the variation in received signal strength due to transmitted energy being reflected back to Earth by an ionised layer in the upper atmosphere. All would be well if the reflected energy arrived in phase with the direct, ground-path energy between transmitter and receiver, but the ionised layer is unstable to the extent that reflected energy varies in phase relative to the direct signal path, sometimes being equal and sometimes opposite, thus causing variations in effective received signal strength.

For security we can turn this problem to our advantage for detecting a man trying to evade the system by crawling under the direct beam. In our example of fading, above, the ground surface at the perimeter can be thought of as the ionised reflecting layer and the crawling man as the instability leading to phase shift of received signals. The resultant 'fading' is just what we need to trigger the alarm. So, by using the standing wave concept together with phase-sensitive detection instead of only amplitude detection, the alarm is raised in spite of the intruder having crawled under the direct beam.

Another feature of interest is that with an elevated aerial and fairly wide vertical beam width the standing wave version of the microwave fence tends to be able to fill in minor undulations of the ground surface.

Unwanted fading Now, if you remember the other problem with standing wave ultrasonics, you will be asking whether this, too, can affect the phase-sensing microwave fence.

It is true that difficulties can arise when no intruder is there, owing to changes in ground reflectivity. These changes may be slow with growing grass, or more rapid in a shower of rain. If the changes are sufficient to affect normal phase relationships more than a preset amount, then a false alarm could be caused.

Fortunately, compared with the smooth walls of a room causing trouble to standing wave ultrasonic sensors, the ground outdoors is rough and undulating, so that very many waves are arriving, each over a slightly different route, and there is only an extremely low risk of sufficient cancellation happening to cause a false alarm.

Yet again, it can happen, and on the larger installations using many microwave sensors something may have to be done about it. You could probably allow a sensor one such false alarm a year, but if you have 52 sensors round a site perimeter, that means on average that the system will give one false alarm a week. If that matters, and I think it does, then can you 'invent' a simple way of solving the problem? For a clue, read Chapter 15 again. This could be a discussion point, and is included in suggestions at the end of this chapter.

Relating wavelength, beam width and detection

An immediate bonus gained from using the microwave beam breaking system shown in *Figure 20.1(a)* rather than using infrared beams stems from the size of the aerial. In Chapter 14 we noted that the detection sensitivity is confined to the 50 mm diameter of the beam between transmitter and receiver. If a bird is big enough to intercept a beam of this diameter completely, a human being also must intercept it completely, and no discrimination is possible between real and false alarm on that information alone.

Using a beam-breaking microwave sensor, the aerial diameter is more likely to be about 300 mm; and a bird is much less likely to be able to intercept a 300 mm diameter beam completely, while a human would be hard put to it to avoid complete interception. Thus, by increasing wavelength and therefore 'aerial' size to retain control of the beam width, the prospects for inherent discrimination between wanted and unwanted targets improve.

Further increases in wavelength on this basis should allow for still greater discrimination, to which is added the advantage that the longer microwave lengths see through paper and sundry non-metallic objects such as leaves, which consequently are unlikely to cause false alarms, as would be inevitable with infrared beams. If one says that subsequent signal processing would prevent a false alarm from an infrared system, one has to say that similar signal processing can be added to the microwave system to extend its rejection capabilities, and the balance of advantages remains more or less unchanged.

The limit is reached for increasing wavelength when the aerial size becomes unwieldy and too big for the space available. In the absence of unlimited choice, owing to Government regulation of usable wavelengths, the practical limit is perhaps about 12 cm wavelength, which allows very reliable transmitters and receivers to be used.

Using shorter wavelengths

In *Figure 20.1(c)* the merits of using phase-sensitive detection are compared with those of the other two microwave fence configurations. Where space is severely limited this configuration has even more possibilities. To narrow the beam width to take up less space, we can either reduce the wavelength or increase the aerial aperture. What happens if we do both? As we shall see, surprises are in store.

Yet another of the scientists of a hundred years and more ago, Fraun-hofer, discovered that in the optical equivalent of Young's experiment (*Figure 16.3*) he could reduce the beam angle to zero if he made the aperture (aerial length) wide enough. Later rather than sooner, of course, as you can visualise from Young's ripple tank, the waves do start to spread out.

Fraunhofer said that an optical beam would be parallel for $2D$ times the aperture divided by the wavelength, where D is the length of the aperture. That is, in a theoretical world, so in the practical world of microwaves let us say that the beam is parallel for only D times the aperture divided by the wavelength, and see what happens.

Suppose that we start with an aerial working at 3.2 cm wavelength (X band) and give it an aperture of 32 cm. Aperture divided by wavelength is 10, or, as we say, the aperture is 10 wavelengths long. Putting this into the practical version of Fraunhofer's equation gives us a parallel beam distance D of $0.32 \times 10 = 3.2$ m – rather an anticlimax. If our microwave fence has to work over a distance of 100 m, this parallel distance is so short that it is always ignored.

By changing the wavelength to 8 mm (Q band), one-quarter of that at the X band, and by changing the aperture to 128 cm, four times as long as our original aerial, the aerial is now 128 divided by 0.8, or 160 wavelengths long. We have only to multiply that by D, or 1.25 m, to find that our parallel beam is now 160×1.25 or 205 m long! This would be long enough for practically any microwave fence application likely to be met, and it would require a 'no man's land' of little more than 1 m wide, with 2 m as a comfortable practical spacing between a boundary fence and an inner security fence.

The advantages of a parallel beam are as follows:

(1) Because the beam is now parallel in the horizontal plane, and spreads out only in the vertical plane, doubling the distance between trans-mitter and receiver only halves the power arriving at the receiver instead of reducing it to a quarter (the ciné screen makes this easy to understand). We can therefore make better use of the modest power allowed by governments for microwave fences.
(2) The detection sensitivity along the beam is much more consistent, because the proportion of the beam filled by an intruder does not change so much with distance from a transmitter or receiver.
(3) The ground space needed for the security system is reduced to the practical minimum.
(4) Nature's fourth-power law, which applies to radar, has been beaten down to a square law by the conventional microwave fence, and beaten again down to a linear law with the parallel-beam microwave fence.

Parallel beams and methods of overcoming deep fades in standing wave phase-sensitive microwave fences are two fundamentals that appear to have been overlooked. If, in fact, these principles have not been applied in practice, who will be the first to use them?

Applications of microwave fences

It is rather a process of elimination finding suitable applications, and applications to avoid.

As we said at the beginning of the chapter, the microwave fence came into being to give early warning of penetration of site boundaries. This can be extended to include the perimeter of one or more vulnerable areas within the site perimeter.

As the line of a microwave fence is virtually invisible, it is necessary to install it within a physical boundary fence to prevent people innocently straying through it. Similarly, a further physical fence, a perimeter fence, is desirable within the microwave fence to prevent occupiers of the site causing false alarms by straying outwards into the microwave fence zones. In critical sites this inner perimeter fence might well carry an alternative detection system, but whatever the detail concept, the idea of a fence within a fence is always liable to meet objections on grounds of excessive cost. In my view, however, unless alternatives can be found, that is a price that will have to be paid if security officers are to have confidence in the perimeter system. Without that confidence the system is not worth having, because a real intrusion alarm can be dismissed in their minds as just another false alarm, and ignored.

One possible alternative to the fence within a fence is closed-circuit television, if it can be used and acted upon quickly enough. The features of this system are discussed in Chapter 7 and 22.

Evasion resistance is a further reason for using a microwave fence. Just as it is difficult for the innocent to know just where the sensitive zone is, it is difficult enough for the intruder even with the two beam breaking systems shown in *Figure 20.1*, but it is almost impossible to be sure of evading the phase-sensing system of *Figure 20.1(c)*, where the vertical coverage can be more than 5 m high.

Even at the transmitter and receiver locations, design problems with aerial 'side-lobes' are put to good advantage in giving coverage where it might not be expected. This is augmented by overlapping the beams at corners and in long stretches of fence.

If the site has too many corners or changes of direction, an excessive number of transmitter-receiver pairs may seem to be necessary. A saving might be achieved by contracting the whole concept and concentrating on the perimeters of the really vulnerable points.

Should the ground undulate more than can be levelled out at reasonable cost, then the microwave fence system is unsuitable and a terrain-following system needs to be considered.

Power supplies and signal communication

When transistor and other high-efficiency microwave sources are used in place of the Gunn diode, the power requirements are reduced sufficiently to make practicable the use of solar cells. A small rechargeable battery at each sensor head can be charged sufficiently during each day by the solar cells to operate the system satisfactorily through the night.

For communication each sensor can be modulated to identify its location, and as the microwave fence is a point-to-point communication system already, it is a relatively simple matter to signal back to the security office whenever and wherever intrusion is sensed. So the cost of perimeter cabling and making good can be eliminated.

Proprietary equipment

Exponents of the various microwave fence techniques tend to follow very individualistic lines, but each is inclined to market a similar low-cost common denominator version.

Taking the techniques in the sequence covered in *Figure 20.1*, an example of the beam-breaking concept is shown in *Figure 20.2*. The dish aerial enclosed in a weatherproof housing gives a conical beam and it uses a gallium arsenide FET microwave source in place of the original Gunn diode which made it all possible.

Racal Guardall in the UK introduced beam breaking with ground cover, and an example of this type of equipment is their Fenceguard Type FG310 shown in *Figure 20.3*. The method used for beam shaping is clearly shown by the vertical aerial, backed by a vertical parabolic cylindrical reflector.

The third type of microwave fence, using phase-sensitive detection techniques, with aerials horizontal instead of vertical, is the product of the UK-based international company Shorrock Security Systems. Their range includes the portable system illustrated in *Figure 20.4*, which can be deployed, for instance, as temporary protection of a parked aircraft.

Figure 20.2 The weatherproof housing at one end of the conical beam microwave fence

Figure 20.3 Microwave fence using a vertical aerial to give ground cover. A pair of these aerials is used for transmit and receive, respectively, enclosed in a radome column cover. (Supplied by Guardall Ltd)

Figure 20.4 Phase-sensitive detection techniques are used in some microwave fencing systems. Here four sets of equipment are shown forming a portable kit for protection of a temporary risk. (Supplied by Shorrock Security Systems)

Discussion points

So many hints have been dropped during the course of this chapter that it is unlikely that finding the solution to the standing wave problem in the phase-sensitive detection system will occupy you for very long. But knowing what to do is still a few steps away from knowing how to do it tidily and economically. For further clues it may be necessary to set up a portable microwave fence, preferably on a concrete hardstanding, and with the transmitter at, say, 1 m high, try varying the height of the receiver and varying the distance between the receiver and transmitter while monitoring the received signal strength. Hopefully, the figures will be sufficient to lead towards debatable solutions. If they are not, can you convince yourself and your colleagues that the whole idea of the standing wave problem is a red herring?

A satisfactory airing of this one topic will reveal virtually all that one needs to know about microwave fences.

21 Field-effect sensors

It has been customary to think of field-effect sensors as proximity devices; consequently, when used in that way, they proved so troublesome that they were regarded as untamable and were discarded. If, instead, they are regarded as interception devices like infrared beams and the conical-beam microwave fence, with the terrain-following characteristics of field-effect techniques, an invaluable weapon is added to the intruder detection armoury.

Electric field sensors

Let us deal with the words first.

'Electric' does not mean electrified; it has nothing to do with the electric fence sometimes used to form a cattle-pen by giving the cattle an electric shock to discourage them from straying. When used for perimeter intruder detection, electric field sensors are more like a capacitor, carrying only a low safe voltage.

If you want to go a little further in understanding how an electric field sensor can work, think first of the capacitor just mentioned, connected in parallel with a coil or inductor to form a resonant or tuned circuit, just as in a radio receiver. You know that by turning the tuning knob the resonant frequency is changed and other radio stations are tuned in. Now, to make an intruder-detector we can make the ground or earth act as one electrode in the capacitor and suspend a wire above the ground a metre or so high to form the other electrode. Back in an equipment box we can connect an inductor across the wire 'capacitor' and with suitable electronics make the circuit oscillate at a definite frequency.

In an electric field system we have to ensure that it is the capacitor in our tuned circuit that is affected, and not the inductor. The capacitance of an intruder is quite small, so we have to make the tuned-circuit capacitance small also, in proportion, so that the intruder can by his presence change the resonant frequency sufficiently to raise the alarm.

To avoid the standing-wave phase-shift effect with movement we have to make the wavelength as long as practicable, and that means using a high value of inductance. This is no problem electronically: it just means that the resonant frequency has to be low. In a fence application, for instance, the length of each detection zone is a major controlling factor.

The shorter the zones, the more precise can be the location of an intrusion, but the extra control and indication equipment can add to cost. Depending upon the manufacturer, recommended zone lengths range between 30 and 150 metres, and corresponding resonant frequencies range from 150 kHz down to the border between audio and ultrasonic frequencies.

We shall now look at typical applications of field-effect techniques.

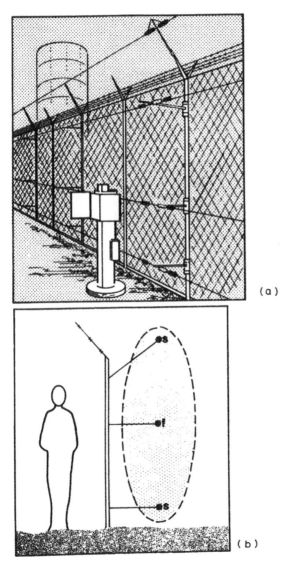

(a)

(b)

Figure 21.1   (a) Perspective view of Stellar E field sensor installed on a perimeter wire fence; (b) typical sensitive area of Stellar three-wire E field sensor

Electric fields used with physical fences

One of the early arrangements is shown in *Figure 21.1*. In practice, instead of relying upon the ground itself to provide one electrode of the capacitor, more uniform results are provided by fitting a further wire to represent the ground. This method has the advantage that a second 'ground' wire can be fitted near the top of the physical fence to provide an electrically balanced arrangement as shown (wires above and below the field wire f in *Figure 21.1(b)*).

An additional practical point is the provision of further wires to help to reduce the spread of the field and so enhance the penetration rather than proximity nature of the detection system. The ground space that needs to be allocated to security then becomes comparable with that needed by stacked infrared beams. Also, of course, movement of people and animals on the other side of the physical fence can go on with negligible risk of causing false alarms. *Figure 21.2* illustrates a fence with electric field sensing.

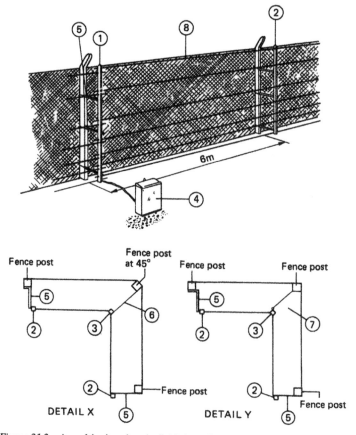

Figure 21.2   A multi-wire electric field detection system fitted to a chain-link fence. The fence mounting posts must be securely fixed in the ground; the fence fabric should be taut; and the T-Line mounting posts and brackets must be secure. (Supplied by Shorrock Security Systems Ltd)

Free-standing electric field detection

For open relatively uninhabited areas, where early warning of approaching intruders is needed, the electric field detection system can be used without a physical fence in its free-standing configuration. Without physical barriers the risk of false alarm from legitimate movement is inevitably higher: hence the likely restriction to early warning rather than full alert.

However, when protection of a high-risk area within a large perimeter is needed, and where even legitimate movement is closely controlled, the free-standing electric field system can be used in the full alert mode. It could be selected, for instance, to deal with rough, undulating terrain, and where the risk of crawling had to be considered.

Roof protection

When reading through this book, you may have felt that disturbingly little has been said about the protection of roofs. So much attention has been devoted to space detection within buildings that, in a way, roof protection is thought to be of less importance, and very few systems specifically designed for it are available. However, space detection within the building is almost the last line of defence, and perimeter detection can give much earlier warning of trouble, before actual penetration of the risk area.

An important application of electric field sensing is to roof access, as shown in *Figure 21.3*. For most roof arrangements it is difficult to see how an intruder can get on to the roof without going through the wires. Detection sensitivity can therefore be kept very low, to avoid false alarms, and, in any case, there is very little up there to cause alarms. Birds are not such a problem as might be thought, provided that the sensitive zone does not cover the ledge where a bird might alight and walk. When a bird is in flight, it is not 'connected' to ground and provides less influence on the sensor capacitance than a grounded object.

Figure 21.3 Using the Shorrock T-Line system for roof protection. (Supplied by Shorrock Security Systems Ltd)

False alarm control

In Chapter 32 emphasis is placed on the need to earth electronic security systems satisfactorily and to check the earthing regularly as part of routine maintenance. With electric field systems, earthing becomes imperative. Without it the required capacitor created by the wires 'floats' like an unwanted capacitor between the wires and the ground. Single-point earthing is also important, normally made at the control unit power supply.

Just as for indoor systems, the equipment itself tends to be reliable, but owing to the relatively large amount of external wiring associated with electric field systems outdoors, interconnection faults are a source of false alarms. I like to see multistrand wire made mandatory, soldering avoided and joints made on the successive lap principle known in the UK as the PO joint. This is very strong, easy to make and reliable.

As with other types of perimeter detection systems, 'gardening' is necessary to trim back overhanging branches of tree, to check weed and shrub growth, and for general tidiness. A less obvious requirement is to check insulators and to clean them from time to time, removing possible accumulations such as salt, sulphur and carbon.

When the first edition of this book was written there was relatively little application of the electric field technique out of doors. Actual experience of a number of manufacturers and users is now well documented, and their reports should be studied in detail for a fuller understanding of the external causes of false alarm and of what precautions to take.

Electric fields for indoor detection

Whether pictures or safes, sports trophies or video recorders, there always seems to be someone who wants to take them from their owner. Many methods of protection have been described but new ways can be helpful in deceiving the intruder. Not that electric field or capacitance detectors are new, but the technology needed to make them reliable was rather late in being applied to them. Tunstall Security Ltd have used the phase-locked loop principle to hold the system stable as slow environmental changes take place, yet removal of the protected object changes the external capacitance of the system and triggers an alarm. Although useful during the silent hours this method comes into its own in museums, art galleries and the like when protection has to be maintained unobtrusively while the public are present. Space detection devices of the PIR, ultrasonic and microwave radar types are useless under these conditions.

Magnetic field sensors

The Stellar H field system is an example of this technique, which consists of buried cables designed to 'leak' radio frequency energy and to detect intruders by sensing the effect of energy reflected from someone walking over the cables.

Principle of operation

If a buried cable is transmitting radiofrequency energy, one can envisage that the energy is going in all directions round the cable, some into the Earth and some of it into the space above. If another leaky cable is buried nearby, it will pick up some of the energy radiated directly towards it. Now, if you remember Chapters 15, 16 and 20, you can be sure of what is coming next. As soon as an intruder walks over the cables, some of the energy going into space will be reflected back from the intruder to the receiver cable. Then, just as with radio fading, some of the reflected energy will be out of phase with the direct energy reaching the receiver cable, and electronic circuitry at a remote control position can sense the effect of the phase shift and can trigger an alarm.

Application to outdoor detection

In a typical installation the cables are buried a few centimetres deep and run parallel just inside a perimeter fence or wall. This can give a detection zone about 2 m wide. If sufficient perimeter space can be allocated to security, the detection zone can be doubled in width by adding an extra receiver cable spaced a metre or so away from the now central transmitter cable.

Contrasting the H field with the E field systems, we now deliberately want the radio fading phase-shift effect when an intruder crosses the cables, so the wavelength has to be very much shorter and typically it is chosen in the VHF band. But it cannot be too short, otherwise the ground will absorb too much of the energy, leaving insufficient to provide the direct link between 'transmit' and 'receive' cables.

False alarm control

Although the phase-shift principle encountered in the fading of long-distance radio reception is the basis of operation, the two problems that killed the standing wave ultrasonic sensor do not apply.

First, because radio or electromagnetic energy is used instead of acoustic energy, the system cannot be put into false alarm by air movement (see Chapter 15).

Second, as the spacing between the cables is so small compared with the cable length (up to 100 m or so), it seems unlikely that any embarrassing standing wave pattern can be set up to cause a fade deep enough to trigger a false alarm (see Chapter 20).

It has to be remembered, however, that the intruder affects only a short section of the overall length of each cable zone and he, too, is incapable of producing a large phase shift or fade. So the usual balance has to be struck in setting the sensitivity controls between certainty of detection and the risk of false alarms. Some aid in this is given by selective circuitry designed to identify movement characteristics and likely duration of movement of intruders relative to other potential causes of alarm signals.

Probably the biggest risk of trouble comes from sudden climatic changes such as rainfall, which changes the reflective character of the ground–air interface, as though a mirror had been put there. The risk seems to have been catered for by burying the cables at only a very shallow depth of 5–10 cm. This means that the difference in signal path lengths between that directly between the cables and that reflected from the rain-soaked surface causes insufficient phase shift to trigger a false alarm.

Discussion points

It is not too difficult to find good perimeter sensors for use in terrains that allow line-of-sight operation. The real problems come when the terrain is undulating and the boundary changes direction frequently. Devices such as the electret cable (Chapter 18) fixed to physical fences are limited to the fence within a fence, otherwise they are exposed to false alarms from legitimate movement outside the boundary. On the other hand, the E field system can be on the inside of a boundary fence and can be adjusted to detect penetration, but not to detect movement outside the boundary, and can follow all the terrain undulations that the physical fence can follow.

It is time to review outdoor detection from what has been described in this book and from any other information and experience you can glean, and to assess whether you think that equipment suppliers can meet all the varied requirements or whether there are gaps still to be filled.

22 CCTV

Evidence of the dramatic increase in the use of CCTV is around us for all to see. The evidence, indoors and out, consists of housings having a characteristic shape and style which people associate with CCTV, with the consequent prospect of being seen, behaving legally or otherwise.

The effect on the minds of most of us of the prospect of being seen, even if we cannot see a camera ourselves, is to deter thoughts of wrong doing. The thriving sales of dummy cameras indicates a widespread belief that deterrence is an effective crime prevention method. This is good, and the stronger we can make that feeling of deterrence, the better.

This book, however, is about dealing with situations where deterrence has failed. Some aspects of the problem have to be excluded to keep within one chapter in one volume, so hopefully someone else will tell you about motorway speed cameras and the like.

Using CCTV in security

Previous chapters have discussed the pros and cons of using various types of intruder detection sensors, the common feature of which is they all give a yes–no indication of whether an intruder is there or not – they give no information about the intruder, and so there is uncertainty or incompleteness in the information passed to the alarm receiving centre, and so in the information they pass to the police.

Anything that reduces uncertainty must be helpful, and CCTV can do just that. Another matter to be clear upon is the normal division of responsibility. With inevitable exceptions, the acquisition and presentation of information is by the security industry (equipment and manned) and the information is used by the police. Before proceeding further, it may be wise to read or re-read Chapter 7 on Surveillance.

As the camera housing is the most familiar part of a CCTV system, we will start with that and the camera sensors, and proceed through to monitors and video recorders.

Cameras

Early CCTV systems were dependent upon thermionic picture acquisition tubes not too dissimilar in operation from, but much smaller in size than,

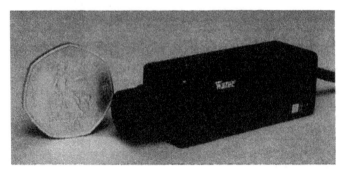

Figure 22.1   One of the smallest 1/4 inch colour cameras available for CCTV.

the domestic TV cathode ray tubes. The 1990's saw these thermionic camera tubes almost totally replaced by solid static charge coupled devices, for both colour and black and white cameras.

The Pulnix Model PE2010 camera is perhaps typical of current technology, with 752 × 582 pixel resolution on monochrome pictures. What is the significance of the pixel rating? It has replaced 'lines per centimetre' used for thermionic tubes and represents one tiny semi conductor-type capacitor which is charged to a potential depending upon the light intensity focused on to it by the camera lens. The capacitor is discharged by the scanning circuitry within the camera, and used to control the monitor brightness at that point.

The camera has a 'target' size of 1/3 inch, coming midway between the 1/2 inch and the more recently introduced 1/4 inch target, which offers 537 × 505 pixels shown in *Figure 22.1*, with a 50 pence coin for size comparison (Vision Warehouse).

Lenses

Much of what you know from home video camera comes in handy here. But the difference is that most CCTV cameras have to work unattended. They are short, therefore, of the advance information gained by a very short focal length lens, fitted with a rapid iris, and mounted on a very fast pan and tilt head – that of the attendant operator.

As indicated below, it is easy to provide a mechanical pan and tilt head for the CCTV camera, controlled by a remote operator. To judge whether this is sufficient, we have to go back to two fundamentals; what we are proposing that the CCTV should do for the user, and how much time is available for the CCTV system to do it.

The purpose is three-fold, the protected premises need to know that an intruder is there, so that they can do whatever is necessary to protect their property, the police need to know for the same reasons, together with a need for as much information as possible to aid identification of the intruder(s).

Regarding time, try to work out just how long an intruder would be visible to you before he vanished behind a building or other obstruction. We now have a basis for drawing some conclusions.

Focal length

The point about time is that even if you were there, with swivelling eyes and neck, the intruder might be visible for less than 10 seconds. If instead of yourself, you have a CCTV camera on a fixed line of sight, the only way to keep the intruder in view is to have as wide a lens angle of view as possible. Wide-angle lenses have short focal lengths which helps them to stay in focus. Even so, the image on the monitor may not be large enough for recognition of the intruder.

That may be a justification for fitting a variable focal length (zoom) lens to 'enlarge' the picture after initial acquisition. Aperture control to adjust for varying light levels is likely to have to be automatic. If the required time on-screen and angle of view cannot be achieved with combinations of the above, the likely answer is to use a combined pan, tilt and zoom (PTZ) head, operated from the local security control room.

Camera housings and pan-and-tilt heads

For indoor fixed line of sight cameras it is unusual to provide a camera housing, the case provided by the camera supplier is sufficient. The exception is when camouflage is needed, and this tends to be custom made to suit the environment.

For outdoor applications a weather resistant housing becomes necessary to enclose the camera and lens completely. With the lens inside, a window becomes necessary, and with the window, the options of a wiper, water bottle, remote controls for the wiper and water spray and so on.

For movement of the camera horizontally (panning) and vertically, a remotely controlled pan-and-tilt head is required. When slow panning of the camera is adequate it is still customary to mount the camera

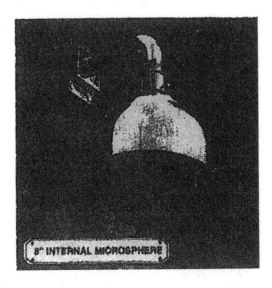

8" INTERNAL MICROSPHERE

Figure 22.2 This head can pan a camera at 200° per second and can include zoom and auto focus facilities.

horizontally. Where fast panning may be necessary, as in a shopping area, the motor torque required to turn the camera is excessive. The solution is to mount the camera vertically, and to reflect the scene into the camera lens with a mirror. Pan speeds of 200° per second are achieved with heads such as the Microsphere, supplied by Video Controls, and illustrated in *Figure 22.2*.

Lighting

Once the principles are established, together with rule of thumb design data for any given type of lamp, lighting for CCTV is relatively simple to handle. However, it is quite another matter trying to start from scratch, and the help of a lighting specialist is as advisable for you as the help of a security specialist is for others.

The main problem is the inverse square law we met in relation to radar (Chapter 4, page 30). You will remember the analogy of the ciné screen; if you double the distance from the light source to the screen, the light covers four times the area on the screen, or is just one-quarter of its previous brightness. As it is reflected light we use for CCTV, just as it is reflected energy we use in radar, that reflected light has to cover the further fourfold increase in area, and the brightness is reduced to a further quarter of the original. A quarter times a quarter is one-sixteenth, a dramatic loss of light for just double the distance away.

If the lighting is to be on continuously after dark outdoors, or any time indoors, any of the gas discharge types of lamp will be more efficient and more economical to run than filament lamps. For security situations where the lighting is switched on only when triggered by an intrusion sensor it becomes almost certain that filament lamps will have to be used. The reason, again, is the time available in which the intruder can be seen. Filament lamps come to full illumination almost immediately, while other lamps take seconds to warm up – time we cannot afford. As filament lamps used in this way are on for relatively short periods, their inefficiency tends to be unimportant. Normal domestic lamps have tungsten filaments, and, in security, filament lamps are more often of the tungsten halogen type, which take an intermediate position in the range of efficiencies.

A rule of thumb used for budgetary estimates of lighting requirements, but not for design, has been suggested by DE Mutch, of Pope Lighting Ltd. To find the number of lamps needed for a given area, first find the total lamp lumens from the following equation:

$$\text{Total lumens} = \frac{\text{area (m}^2) \times \text{required lux value}}{0.23}$$

Each lamp has a lumen per watt and a wattage rating or total lumen output rating, so the number of lamps can be estimated by dividing total lumens required by lumens per lamp. (The lumen, (lm) is the unit of luminous flux, or light energy.)

Mounting height is judged in conjunction with light-spread angle to avoid dark patches on the ground, and the lux value required from there. If lighting is required also to dazzle an intruder, it will have to be mounted quite low and to look outwards. Thus, some effective light may be lost, and further advice is needed. For zero light infrared lamps, it is best to seek the advice of the infrared camera supplier.

Picture transmission

Once cameras and lighting having been decided upon, decisions are needed on how to link to the monitor. The decisions turn almost entirely upon the distances between the cameras and the monitors and on the environmental nature of the route in between.

Twisted pair

For short distances there is technically little against using a pair of wires such as telephone lines, low voltage d.c. supply wires, alarm signalling and control circuits, anything, almost, that happens already to be part of the system, provided that the TV equipment is suitably isolated with series capacitors. Indeed, in the early days of television outside broadcasts, the public telephone cable network had to be used in this way by broadcasting authorities – there was no other way.

Two major snags limit the use of telephone wires, often referred to as twisted pairs, and similar two-wire circuits. One is electrical interference due to inductive and capacitive pick-up by the unscreened pair of wires, although some forms of interference are cancelled out through being at equal and opposite amplitudes in the two wires. The other snag is loss of picture detail due to the relatively high capacitance between the individual wires in each pair. As you know, TV pictures need signals up into the megahertz range of frequencies to give good detail, and the higher the cable capacitance the lower the by-pass impedance in parallel with the wires. Attenuation of the signal, therefore, increases with frequency, and increases for any given frequency with the length of the wires. This can, and had to, be compensated for every so often with equalising filters and amplifiers, but as a result the cost goes up, and other ways are needed.

Coaxial cable

Coaxial cable has a central conductor surrounded by solid or cellular insulation, and well spaced from the central conductor is an outer braided or helically wound tape conductor which acts also as a screen for the inner conductor. Attenuation of high frequencies is much less than with twisted-pair cables and picture transmission over cable lengths of the order of a quarter of a mile are practicable. For that distance a signal loss of about 10 dB may occur at 3.5 MHz, an acceptable upper frequency limit for CCTV work.

Beyond this distance, amplifier equalisers may be necessary, as supplied by various equipment manufacturers, to restore picture quality to that needed at the monitor end of the cable. However, where in a multicamera installation the picture information from several cameras follows a common route to the control room monitors, it may be preferable to use a multiplexed glass-fibre optical system.

Radio link

Situations arise where it is impracticable or inconvenient to lay or suspend a coaxial cable from the camera to the monitor. In that case at least part of the transmission link can be by radio.

The aerial of a radio transmitter is set up at a suitable height to see across the difficult terrain towards a receiver aerial set up on the monitor side of the site. Instead of broadcasting the picture in all directions, it is normal to use very directional aerials to give, in effect, only a pencil beam of information between transmitter and receiver, operating in the microwave bands allocated for links. The dish aerials used for high directivity conserve transmitter power, improve security and reduce the risk of deliberate interference by jamming. Coaxial cables are normally used to feed the respective transmitters and receivers.

TV Monitors

In Chapter 7, reasons were given for discouraging the use of more than, say, three monitor screens in a security control room. This applies to nearly all situations except where the motion alarm sensing display system is used, when rather more monitors are acceptable.

Of the three monitors recommended, it seems preferable from the boredom–inattention point of view to have no picture at all on two monitors, until an independent intruder detector triggers. Once the sensor has triggered, the same signal can cause the camera overseeing that area to be switched through to one of the monitors. In order to get the security officer's mind on to the subject with minimum delay, it is usually necessary also to have a signal light indicator associated with each monitor to say, in effect, 'Look at me', as well as having the normal audible alarm sounding.

A form of switch that has a valuable place in addition to those mentioned above is the sequential scanner. If connected to the third monitor, the switch can bring the scene from each camera to the control room for a few seconds, changing automatically so that the whole risk area is seen once a minute or so. This is of little use for intrusion detection, but it proves that each camera is working, and it helps to ensure that the security officers become familiar with the site as seen via CCTV. The sequential scanner switch can be over-ridden manually to hold any scene for a longer period if needed.

The blank screen

The advantages of having blank monitor screens also include the avoidance of 'burning-on' a static scene on to the screen, which can impair the contrast of a scene when it changes suddenly.

For security, loss of contrast is significant from another aspect. If you are reading a printed page, and you suddenly look up at the TV screen or at someone who has just walked into the room, your eyes have to change focus to make up for the change in viewing distance. The eye knows that it is in focus when the contrast is at a maximum. So with security CCTV monitors, if the security officer suddenly has to give his whole attention to the screen, his eyes will adjust more quickly the better the picture contrast.

In choosing TV monitors, good focus and good contrast are important criteria and are best judged by comparing different makes and models. However, do not be misled by the apparently superb pictures on very small monitors. Much of the detail that seems to be there is supplied by some trick of the eye akin to imagination. The test is whether you can extract detail information quickly from a changing scene from the normal operational viewing distance. This test is likely to lead you to choose a screen size rather larger than you initially thought to be attractive.

Video motion detection

A valuable alternative to the blank screen for avoiding boredom problems is a technique called video motion detection. The system is quite simple in concept, quite complex in design and circuitry, and quite easy in application. Here we need concern ourselves only with the easy bits.

In concept, video motion detection acknowledges the problems of human nature and physiology, and says, in effect, 'Let us do something to attract the attention of the operator to a monitor showing an incident.' In operation, a motion alarm system can look much the same as any other multiple camera and multiple monitor TV surveillance system. The difference is that circuitry is introduced to sense when the picture changes. The picture from a camera looking at a fence cannot change much unless an animal or a person appears on the scene. Once that happens an alarm signal is triggered automatically to attract the operator's attention to the monitor showing the intrusion. The operator really would have to be negligent to ignore these warnings, so operationally and in concept the video motion detection principle is sound. The equipment is available from various suppliers, including, for example, the Video Tracker from Primary Images Ltd.

Identification

From the police point of view, as users, the ability to identify a person as seen on a monitor screen can be paramount, and the whole CCTV design and equipment must be geared to this need. Unless the camera can provide the information, nothing else in the chain can retrieve the situation.

Clearly, to achieve recognition depends upon how far from the camera (fitted with an appropriate lens) the target person is at the critical time. The nearer the target, the larger is the image on the monitor screen, and the lower is the resolution or definition needed to achieve identification. Monitoring of access control and keeping watch over an open field are examples of easy and difficult tasks.

If you say that a face occupying one-tenth of the monitor screen height must be recognisable, you have established in general the definition standard required of the whole system, including the police video and monitor, and of the camera in particular.

To discover whether equipment exists to meet this standard, or to check whether the camera is, after all, the limiting factor for instance, you will need much more information than can be given here, but references are given at the end of the chapter that could be helpful.

Video recording

It is the first few seconds after a camera has been triggered through to a monitor that is crucial in using CCTV to help filter out false alarms and to help the security officers to react correctly to a real incident. And yet it is those same few seconds after perhaps months of 'nothing happening' that a security officer's concentration is not at its best.

One would have expected even so, that after such a wait not only would the incident have been noticed and acted upon, but that it would have been recorded with sufficient detail to be of value to the police. The situation is summarised somewhat by a saying, such as, 'A good camera caught him, but he was let off by an inadequate recording.'

The recorder is the last in the chain of electromechanical equipment and is the interface between the security industry and the police – who need evidence.

Has the industry done enough? In the second edition of this book I answered this question by suggesting an improved type of recorder. The suggestion proved sound in principle but lacking in practice, because no tape existed capable of recording detailed information when used in a continuously running endless loop. Since the second edition, digital everything has taken off, and recorder suppliers, scared perhaps of being left behind, switched their development effort to digital recording, rather than completing their work on analogue systems. It seems that digital systems may not overcome economically the problem of providing adequate storage capacity. One solution to that problem may be to take the principle of my original suggestion but instead of using the endless loop of tape needed for immediate recording of information from the camera, we might use a digital recording head with less than a minute of storage capacity, then convert to analogue for conventional storage. If they are satisfied that not too much picture detail is lost in the conversion, someone may already be doing something like this. No matter, so long as the overall problem of providing satisfactory recording equipment is solved.

Discussion points

If you did not exhaust Chapter 7, there may be controversial points from there that you could continue here. For a lock-up-and-leave site, for instance, would you recommend using CCTV cameras feeding a blank screen monitor, or should there be multiple monitors with one-screen motion detection to raise an alarm?

If you could persuade your Trade Association to organise a poll of members on what features your ideal video recorder would have, what would be your recommendations?

The Aldridge Papers

For an advanced briefing from the user's point of view on the detail operational requirements for CCTV, there is probably no better source than a series of articles written by Mr J Aldridge of the Police Scientific Development Branch.

The Aldridge papers, if I may call them that, cover:

(1) *CCTV Operational Requirements Manual*
   Publication No. 17/94 Police Scientific Development Branch, ISBN 185893 3398. Effectiveness programme.
(2) *Performance Testing of CCTV Perimeter Surveillance Systems*
   Publication No. 14/95 ISBN 18589 35369.
(3) *Effective CCTV*
   Security and safety systems.
(4) *Operational Requirement Analysis*
   A new approach to effective security.
(5) *Town Centre CCTV*
   Good practice check list.
(6) *Video and Image Quality*
   Relating picture detail to each system component.

The Aldridge papers are under controlled circulation. Application for copies may be made to: The Information Service, PSDB Woodcote Hill, Sawbridge, St Albans, Herts AL4 9HQ, UK.

23 Doors and door control

In Part 1 of this book an effort has been made to encourage understanding of a subject rather than learning by heart, and in particular to encourage understanding of things that cannot be seen, such as infrared, ultrasonic and microwave detection. This has led to a fairly full discussion of how these devices work, before presenting examples of the devices themselves in Part 2.

We have reached the stage now when most of the 'how it works' discussion has been completed, and the remainder of Part 2 concerns equipment that relates more to everyday life and can more readily be understood by seeing and using it. However, there remains always the question of good and not so good ways of using equipment, and comments on design and user techniques will continue to be made in the chapters that follow.

Chapter 24 will deal with signalling and control equipment and Chapter 25 will cover access control. Superficially, that may seem to be control of yet more equipment, but further thought shows that we are getting towards the purpose of it all – the actual control of people.

What is a door?

The dictionary definition is very literal, and reflects perhaps the general lack of awareness among people of the function of a door as a security device. If we have something to protect we think of a box. The larger the object, the larger the box – a safe, a strongroom, a vault, a Fort Knox. For real security the top and bottom and all four sides need to be solid and strong enough to resist all forms of attack. That may be acceptable for nuclear waste but it is no use to us if we need access to the object ourselves sometimes. If we provide a hole in the box we destroy its security value, unless we can block the hole.

The dictionary mentions a hinged or sliding panel, but it says nothing about the quality of the door nor about the bolts and locks needed to help restore the security value. That is why this chapter was added to the second edition of this book, to emphasize the importance of door design and door control in security.

Doors

The main types of door encountered in security are listed and explained in Chapter 10. There they were treated in the context of access control. There are far more doors that are not involved with access control but the security principles remain the same.

If we try to establish a standard by which other doors may be judged we might start with the body of the door itself. The standard would probably have to resist attack by ram, axe, hammer, torch, firearm, drill, saw or explosive. Those concerned with resisting terrorism often favour a highly compressed multi-ply and resin-bonded wooden construction. Probably the best example of this concept is known as Delignit, produced in Germany. The UK agents are LT Lewis and Co., of Chingford. *Figure 23.1* shows a section of a test sample of Delignit after a bullet had been fired into it. Given that this is a fair reference-standard material, then, depending upon the degree of risk involved, decisions can be made on relaxing to a lower standard or requiring a higher one.

The frame into which the door is fitted is equally important, and for a reference standard could be machined from the same material, fixed into the wall aperture with long, stout rag bolts.

Door furniture is considered next, and *Figure 23.2* is a useful guide. The pivot 'pins' in hinges are a favourite target for villains, and the illustration

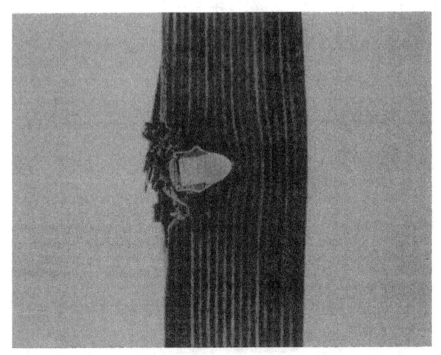

Figure 23.1 A 9 mm bullet fired into 30 mm thick Delignit from a distance of 3 metres results in less than 50% penetration. (Supplied by *Security Gazette*)

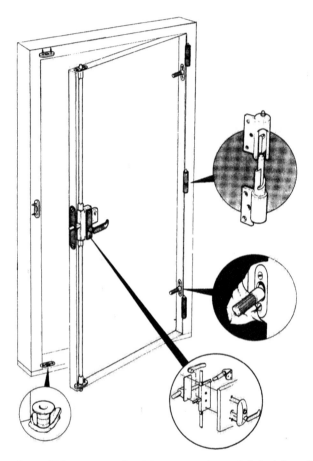

Figure 23.2   An example of the construction and fittings for a final exit door, with options when used also as a daytime personnel door. The bottom right inset shows the 'locking with emergency override (LEO)' mechanism. (Supplied by Surelock McGill Ltd)

shows in the top inset a type of hinge in which the pin is capped and inaccessible at both ends. Attack by ramming is resisted by hinge bolts, shown in the centre inset. Finally, bolting and locking of the swinging edge of a door is crucial; in both a reference standard and in situations of real security, bolting in the centre is inadequate. The illustration shows triple bolting, at top, centre and bottom. All hardware mentioned is supplied by Surelock McGill Ltd. Such a door set is suitable for use on its own, or, with the addition of a solenoid to the bolting mechanism, it is compatible with most proprietary access control systems.

Door control

In order to crystallize the concept of door control it is worth repeating a sentence or two from Chapter 10. When you purchase a key to control a door you can hardly avoid buying a matching lock with it, so, in one act, you obtain both the mechanical part for holding the door closed (the lock) and the means to operate it (the key).

On the other hand, if you go out to buy access control equipment what you get in effect is only the key. Quite separate decisions and purchasing actions can arise when it comes to getting the door bolting and locking mechanism – for door control.

Single-leaf doors

These are customary as personnel doors and are so taken for granted as such that one often finds access control systems being added to existing personnel doors. There seem to be no satisfactory countermeasures to evasion of the single-leaf door by holding open, leaving open or blocking open, and the only likely way of using them in an access control system is to add equipment which will alert the security personnel that evasion is taking place. One way is to use a door contact with a time delay adjusted to expire and trigger an alert after the door has been open long enough to allow one person to go through, but not long enough to allow two people to go through. There are clear risks of unnecessary alerts being caused by such a system, and a better way is to use not less than two pencil-beam infrared rays across the doorway. If the beams were inter-cepted more than once per access authorisation, an alert would be raised, from which a search for the intruder could be initiated.

All personnel authorised to use the door need to be made aware of the problem, and of the necessity for ensuring that no one follows them on their own access authorisation and that the door is closed behind them before they go on their way.

Much can be done to ensure that the door stays securely closed against an intruder. An automatic door-closer covers the risk of the door simply being left open through carelessness or intent. That, of course, on its own is insufficient. The least that should happen is that the door should latch so that it cannot be opened from outside. Surelock, who specialise in the many aspects of door control, arrange that, as the door closes, top, bottom and centre bolts are released to secure the door into its frame, and the access control solenoid is operated automatically to lock the bolts in the thrown position.

Air-lock doors

Although we have nothing to do with the control of air, the security indus-try has borrowed the term 'air-lock' to signify the same arrangement of doors used in the control of air. The arrangement consists of two doors in sequence, as in a corridor, but so interconnected that as one enters the first door it has to be closed and locked before the second door can be unlocked and opened. The advantages to security are of considerable importance, and nearly all the problems of an unattended single-leaf door are overcome.

With air-lock doors it is no longer possible to leave a door open, accidentally or deliberately, and if two people try to go through together on one authorisation, willingly or under duress, the system can detect the presence of a second person by the infrared beams mentioned earlier, or by a load-cell type of weighing device in the floor between the doors. Once the presence of a second person has been detected, the information can be used to lock both doors, with the people inside, until security personnel can deal with the situation.

The same type of doors as for single-leaf door access can be used for air-lock access, with the addition of a little extra electronic control. The disadvantage to the authorised user is the delay in transit, and this leads to another door arrangement, which is similarly effective in resisting evasion and yet allows reasonably rapid movement.

Turnstiles and revolving doors

Turnstiles are much better than they look at first sight, since they are reasonably quick to use and evasion can be detected readily by visual or electronic means for local or remote security personnel. Turnstiles are probably at their best in conjunction with access control systems at site entrances manned by security personnel. *Figure 23.3* shows a typical turnstile, manufactured by Gunnebo Mayor Ltd.

At unmanned entrances, at the perimeter and within the site, the most secure door control system has to be the revolving door of the solid or

Figure 23.3 Turnstiles are effective aids to security personnel stationed at site entrances, and impose minimal restrictions upon authorized entry. Illustrated is the slimline model. (Supplied by Gunnebo Mayor Ltd)

Figure 23.4 Each of the two Tubestile type TS entrances shown has two curved sliding doors, linked on the air-lock principle, so that one door has to be locked closed before the other door can be opened. The locking can be linked to access control systems as required. (Supplied by Gunnebo Mayor Ltd)

bar type. Evasion is almost impracticable, but the size, needing double the width of other systems, and the relatively high cost, tend to limit its use. There is also some authorised user resistance on account of the rather oppressive appearance. These negative features have led to the evolution of redesigned versions specifically for use with access control systems. That made also by Mayor Turnstiles Ltd (*Figure 23.4*), for instance, will fit as a replacement in a normal-width doorway; Tann Synchronome were also one of the first to realise the need with their type of revolving door.

No type of door is perfect in all respects, but if the needs of the user are taken into consideration with the type of risk and the efforts an intruder would make to get at that risk, a reasonably acceptable solution can be arrived at.

No one should underestimate the number of questions and answers that arise in the detailed design of a door control system for a given site, and this work has to be reflected in the overall price. Without it, money spent on access control equipment is probably wasted.

Discussion points

Next time you go to a site having access control, look at the doors being controlled. Do the doors themselves and methods of keeping them closed match the risk and the standard of access control equipment provided? If not, what lessons can you learn from the situation? Discuss with colleagues what each of you would like to see done about it.

24 Signalling and control equipment

The general principles involved in providing signal control facilities at the risk premises, and the steps needed to convey alarm information to the police or other reaction forces, were outlined in Chapter 8. Here we concentrate more on the methods and equipment needed.

As the requirements of signalling and control differ in the three stages involved – sensor to local control, local control to alarm receiving station and alarm receiving station to police, the three stages are treated separately below. At all stages, the influence of the stringent ACPO anti-false alarm conditions, whether reiterated or not, is evident. Regarding interpretation, the Codes of Practice NACP 14 and SCOPP 103 issued by the Association of British Insurers are helpful.

Sensor to local control

For decades it seemed, all connections between door contacts, infrared beams and space detection sensors have been with the multi-strand four-core cable specially made for the alarm industry. No doubt it will continue to be so used – and why not? As a material it is reliable, inexpensive and easy to use, and it can be unobtrusive. However, the costs of labour for installation of cable can be high, particularly when it is installed high up, out of reach from floor level to minimise tampering or accidental damage. Alternative links have therefore been developed. The breakeven point between cable and other methods can be assessed quite easily, and with time it is likely progressively to favour the radio link.

Radio link

Instead of using cable to pass normal/alarm information from a sensor or a group of sensors to a local control panel, a very small low-power radio transmitter unit is fitted at the sensor end and a corresponding radio receiver is sited at the control panel. Power consumption of sensors and transmitter can be so low that battery operation is practicable. Had it not been, the cost of power cabling could equal or exceed the cost of conventional four-core cabling.

The days quickly passed when it was necessary to purchase a separate transmitter to add to your alarm sensing device, and it is customary to purchase the sensor ready fitted with a radio transmitter. Likewise the control panel, such as the Menvier type TS690R, comes complete with radio receivers compatible with the main makes of transmitter available. Owing to the very low power requirements of typical sensor-transmitter combinations it is readily possible to use them with small PP3 type dry batteries, replaceable say, once a year.

Mains signalling

The electricity supply mains wiring has been in use domestically for many years for linking a nursery with a living room as a 'baby alarm'. Normally it is switched so that a call from upstairs can be heard downstairs, but mains signalling can also be used as a two-way loudspeaking telephone. A few firms have introduced modified versions of the above for signalling from two or more intruder detection sensors to a local alarm control point in a house or small business premises.

In its simpler forms the method is prone to electrical interference and to false alarms, and may not be considered for serious security projects. Further problems include the ease of malicious or accidental disconnection,

Figure 24.1   The Menvier control panel has 30 radio (wire-free) channels for use with sensors in such devices as intruder alarms, panic buttons and smoke alarms

simply by removing the plug from the wall socket; also, the connection is usually between the live or line mains wire and the neutral wire. This is satisfactory so long as the transmitter and receiver points both use the same mains phase line, but in some buildings each floor is wired alternatively to any of the three mains phases and simple systems cannot communicate between one phase and another.

Not on grounds of installation cost but rather on grounds of portability for emergency installations, mains signalling equipment has been developed for communicating between mains neutral and earth. The neutral and, of course, earth are usually common to all phases, so that communication can be maintained between units on different phases, provided that all three phases and neutral are connected to the same electricity substation transformer. At the cost of some complexity in circuitry, high immunity to interference can be achieved and individual transmitters can have their signals coded for location identification. Nevertheless, little future is seen for mains signalling in normal security situations.

Ultrasonic and infrared links

These only need a mention, since both methods are fully developed for remote control of various domestic radio and television receiver functions, and can be adapted for, say, links between sensors and control panels as soon as someone sees the need.

Local control equipment

In Chapter 5 (page 40) reference was made to the high incidence of false alarms received by the police at the locking-up and opening times of protected premises. This problem was attributed to human nature, and to the system design and equipment design of the system control equipment on the premises – the local control equipment. A sound principle in remedial action is to 'remove the cause'. In this instance we cannot remove people, but we can take away the local control equipment, and this theme is developed below.

System control panel

In our business the idea of control involves the decision as to whether to have the detection circuit on or off, and involves the means of implementing that decision – a switch. A typical intruder detection installation incorporates a fair number of detection devices, located in various parts of the building; some parts may be empty of people and needing protection, while others may still be occupied and not needing alarm protection. That is a possible operational situation, which means that more than one control switch is needed, to select those areas needing the alarm to be on from the occupied areas.

From an engineering service and maintenance point of view, allowance has to be made for possible faults arising in detection equipment and links back to the control equipment. Switches are normally incorporated to isolate faulty equipment so that the remainder can be used as needed.

Unless legislation or good design decrees otherwise, there is a further switch, a master-switch which routes all detection devices that have been switched on, through to the security control station.

Already the control panel has acquired several switches, and the potential for causing trouble is clear. Apart also from these problems is the problem of how the last person is to get out of the building and lock up either without causing a false alarm or without omitting to protect his exit (and the villain's entry route).

An expedient adopted by many alarm companies, and even written in to some standards, was to incorporate a time-delay unit into the control panel. With this addition, the last to leave the building could switch on the master-switch and then make for and leave via the final exit door. The duration of the time delay was selected to give him adequate time to get out of and lock up the building before the time-delay switch completed its cycle and automatically switched the detection system through to the police-calling system. The consequences for the police of the last to leave going back for his umbrella were illustrated in the answer to Question 7 in Chapter 5.

Another unhappy device is the microswitch fitted into the lock of the final exit door. The purpose of the microswitch is to include a protection device such as an ultrasonic sensor and perhaps a door contact into the intruder detection system after the last to leave or keyholder, let us call him – has left the building, closed the door and locked it. The snag comes when the keyholder is unsure in the dark which of the many he has is the right key for that door. What is more natural than, with the door still open, for him to try various keys in the lock until he finds the right one? And as the right one throws the lock-bolt with the door still open, the microswitch operates and the stage is set for another false alarm.

Alternatives to the system control panel

Improving security involves improving user discipline but, as we have found in Part 1, human nature can take over from discipline, and this has to be allowed for in design. If we can do away completely with the control panel, all we are left with is the need for a switch at the final exit door to switch the system on as the keyholder locks up the building. But not with a microswitch in the door lock! The system requirement of this concept is clear enough, so it is up to the equipment designer to find ways of satisfying the requirement.

There is much to be said, for instance, for incorporating the system master-switch in the keep or recess in the door frame into which the bolt of the lock must travel to secure the door closed. There is a choice of magnetic methods of switching that are highly tamper-resistant. An argument against using the keep in this way is the risk of the door being jemmied open. If this risk is real, there are two answers at least. One is to use a lock with a long bolt-throw designed to meet that particular challenge, or to use a combined bolting and locking mechanism that secures the door at the top and bottom of the door frame as well as at the side, as explained in Chapter 23. The latter method has the further advantage that a time-controlled solenoid lock can be added inexpensively to the bolt mechanism to ensure that not even the keyholder, or a villain who has managed to get the key, can unlock the door except between predetermined times. If you are taking security seriously, there is no way you can avoid providing a decent door for the final exit.

Two other arguments have to be answered. Without a control panel how can parts of a building be protected when empty while other parts of the building remain occupied? The answer has to be that any unoccupied part of the building ought to be sealed as securely as the whole building, and therefore it should have its own final exit door, equipped as for the building final exit door. Finally, if it was useful to have a control panel for isolating faulty sections of the detection system and to aid maintenance and servicing, it is as well to re-read the insurance policy, where it almost certainly says not only that you must have a security system but also that it must be used. If a section is switched out and intrusion occurs in that area, an insurance claim may be disallowed. Much better, then, not to be able to switch a section out, but to employ an alarm installation company that can give rapid attention to service calls and can repair faults while the building is given temporary cover by attendance of its own staff.

To help service engineers to know which section of the system needs attention, there is every reason for retaining the indicator features of a control panel. Such a fault indicator panel would be used also by the keyholder to check that all sections of the system were 'set' correctly as he went on his locking-up rounds. If he had forgotten to close a door, it would show on the indicator panel. If the designer had thought it through, he would also have provided an automatic solenoid-type interlock to prevent the keyholder from switching on the system at the final exit door so long as the indicator panel showed a fault. If the indicator panel were not near the final exit door, then a simple auxiliary repeater red/green indicator light would be sufficient for the keyholder to see when he was about to lock up. The interlock would prevent him from locking on to a fault, and the range of false alarms related to this routine would be eliminated. In the premises where these principles have been applied the procedures required fitted very closely with the procedures the keyholders were used to.

To some alarm installation companies the concept of abolishing the local control panel may not fit in too well with their own procedures. The control panel tends to be their focal point, and even if they buy in all other security equipment, they tend to manufacture their own control panels, to give a sense of individual identity and a means of instant recognition that it was they who provided the security system. Hopefully, they can transfer their attention and establish their identity on impressive-looking indicator rather than control panels.

Links between local control and alarm response personnel

Once the final exit door of the premises has been locked closed, provision has to be made for communicating an alarm condition to the reaction forces. Although there are several options, neither the customer nor the security company is necessarily free to choose the method to be used. The requirements of both police and insurers have to be met, and for the larger risks, particularly on sites having their own round-the-clock uniformed security personnel, the decision may be also to have their own central control facility.

The bell alarm

The local bell is intended to be an acoustic link between the detection system and an intruder to warn him to go away. It is also an acoustic link with any police officer who happens to be near and with any member of the public who cares to notify the police. Owing to the potential for nuisance to neighbours, the ringing time and the bells themselves and the type of sound they emit are often covered by National Standards, although the style of box into which they are fitted outside buildings is left to individual alarm companies.

Not always, though. There was the classic case when the worthies of a town met with the intention of standardising the appearance of burglar alarm bell-boxes in their town, and they invited the various alarm installation and manufacturing companies to submit samples of their wares, from which the approved version was to be selected. Bell-boxes are not designed just as covers; they are designed also to resist attack by villains. Notwithstanding that, the worthies selected the one and only box there that could readily be lassoed and pulled off a wall by a villain.

Automatic telephone dialling

The second option involves methods of using the telephone network. The UK police will no longer accept calls direct from protected premises, but the auto dialer still represents an economical connection link. Instead of dialling the 999 and the police service, the auto dialer dials the number of the security call answering service, run usually by an alarm company. Having called and been answered by the alarm receiving station, the machine then gives a message such as 'suspects at' – then the address of the protected premises. The receiving station then has the opportunity of filtering the call to eliminate possible causes of false alarm before passing the call to the police for action.

Direct line to alarm receiving centre

By far the simplest, quickest and most reliable system is the use of a rented dedicated direct telephone line (private wire) between the premises and an alarm receiving centre. Once the protected premises go into an alarm condition, any of a preselected range of electrical signals can be transmitted on the continuously connected telephone line between the premises and the alarm receiving centre. More information to aid verification can be transmitted by the direct line than by an auto dialer, but once he is as certain as possible that the alarm is real, there is little more information he can give to the police than would be available from an auto dialer.

It has to be said, however, that not all false alarms are caused by faults at the protected premises. Some such alarms are caused by telephone line maintenance engineers who may inadvertently connect test equipment to the telephone line, which has to be made sensitive to such actions because the interference may have been caused by malicious tampering with the telephone line, and tampering has to be treated as a full alert.

Proprietary alarm receiving centres

It has long been the practice of the larger security companies each to have their own 24-hour manned security control room to which their customers premises could be connected.

With the direct line to the police the police accepted whatever calls came in, but had no facilities for querying them. Central station facilities were sold on the basis that if, for instance, a certain customer did not lock up its premises within a previously agreed time from normal closing time, the central station personnel would contact the premises to find the reason why. The reason might be acceptable or it might be found that the keyholder was under duress, and unable to lock up. The police could then be called to attend, even though the alarm had not triggered.

In such ways, the alarm centre personnel can introduce both an element of filtering of false alarms and an element of additional security. Indeed, the use of alarm receiving centre, or Central Stations as many still call them, has become a mandatory requirement in the ACPO policy document. A further requirement became mandatory in January 1997. This requires that all installed signalling systems must be able to transmit either a set/unset signal, or a 'mis-operation' (abort) signal to the alarm receiving centre. If, for instance, following an alarm signal sent accidentally, the alarm receiving centre must be advised in time to prevent them passing on a signal to the police which by then is known to be false.

Private central stations

In situations where the risk is large enough and the premises are large enough, there may be justification for a company to have its own central station on the site, operated by its own security personnel. As with the proprietary central station, the link with the police is normally by a continuously connected direct line. The merit of this system is that firms going to these lengths for security understand better than most the problems involved and are likely to do what they can to filter out false alarm situations. With this background the police can have greater confidence than normal that a call from such premises is likely to be a real alert.

Multiplexed systems

With the cost of renting direct telephone lines in mind, increasing numbers of potential users and continually improving technology, the way is clear for increasing the availability of shared telephone lines and equivalents, dedicated to security and like services. One or two firms have been operating multiplexed alarm signalling links, which allow the shared use of telephone lines, for many years, and the BT RedCARE system is an example.

Dual communication

To conclude this outline study of alarm signal communication and control we must look again at what are the weakest links in the chain – the communication link between the protected premises and the alarm receiving centre, and between the alarm receiving centre and the police. A break in either link can negate all the good work put in elsewhere. A simple, and not too expensive, answer is to use two links in parallel for each leg, preferably using different technologies.

CSL (Communications) Ltd is a firm specialising in providing solutions to the above problem. They say that their Dualcom system combines radio and telephone signalling paths to give continuity of communication. If one signal path fails, either accidentally or deliberately, the functioning path will transmit this event to the alarm receiving centre. If no subsequent signals occur, this alarm can be treated as an engineering fault and the police need not be notified. However, if subsequent signals are transmitted, it is likely that an intruder has entered the premises. With this system, false alarms due to line faults are eliminated.

Discussion points

If trying keys in the lock of the final exit door can cause a false alarm, what would you do to avoid it, and could your precautions expose the system to greater risks of tampering by a villain?

Discuss the merits and otherwise of designing security systems to operate without the use of local control panels. Are you satisfied with the range of signalling systems now available?

25 Access control equipment

From the analysis worked through in Chapter 10, it has to be accepted that access control systems cannot meet the ideal requirement of eliminating the need for supervisory attendants. However, by discriminating automatically between wanted and unwanted people access control enables supervisory attendants to concentrate on the exceptions (the unauthorised people) – and that is a good and acceptable management principle.

A distinction was made in Chapter 10 between the functions of access control and door control, and in considering applications and the suitability of equipment a distinction is needed also between site or building entrance control and movement within the site or building. Breaking these distinctions down further, access control covers establishing identity and the equipment that passes the automatic admit/reject decision to the door control system. Door control covers the locking system, which implements the admit/reject decision, and the bolting system, which physically holds the door in the closed or reject position, or releases to allow the door to be opened.

Identity

If security manpower are charged with the duty of dealing with the exceptions selected by the access control system, the implication is that the majority of people who are likely to seek access are acceptable; and if they are acceptable, they should be admitted automatically. If they are to be admitted automatically, they need some means of proving to the system that they are considered acceptable. This they can do in either of two ways. One involves some individual characteristic of the person, such as his fingerprint; the other, some token given to him, such as an identity card.

The key

Of all identity tokens, the key must be the best known. Possession of a key says that the holder is authorised to use it to gain access to certain premises. That this could be a lie is a weakness. The key could have

Figure 25.1 One of the few early all-mechanical non-key locks for personnel doors. All authorised users use the same code (The Simplex Lock, supplied by Abloy Security)

been found, stolen or obtained from the rightful owner under some form of duress. To this weakness must be added the risk that a duplicate might be made and the original returned without the owner's knowledge. Nevertheless, the key is widely accepted for local control of access.

Combination lock

The main alternative to keys for local control is the combination lock. This device is particularly well known for the doors of safes and strong-rooms, but perhaps less well known for the control of personnel doors is the much simpler, push-button coded door-lock marketed by Abloy Security and illustrated in *Figure 25.1*. The 'token' is the sequence of figures which have to be memorised for operating the push-buttons to release the lock. All people authorised to use that door use the same group code. The advantage is that the 'key' cannot be lost or taken, but without added visual shielding the punching sequence can be memorised by an unauthorised person nearby watching authorised people enter. It works satisfactorily indoors, but on outer perimeters rough usage by people who may not want it to work tends to limit its life.

These, then, are examples of the two significant local and direct mechanical means of establishing identity for access control. Practically all other methods are indirect electromechanical means to be used in the locality of the door.

Keyboard identification

This provides the same protection against loss by theft, etc., as the push-button type of combination lock, and it has the further advantage that, being an electromechanical device not directly coupled to the locking system, it can accept individual identity as well as group codes. Each authorised individual is given and is expected to memorise rather than write down a multifigure code which is entered on to the keyboard, as with a typewriter. When this type of keyboard is installed outside high-risk areas such as banks, adequate care seems to be taken to conceal from unwelcome view any manual entry on the keyboard.

What really bothers me about keypads as normally offered for security is the ease with which an onlooker can read the code entered by a legitimate user. I was trying out an access control/door control system at a busy exhibition, and, having been given a sample code and card, I thought that I had succeeded in proving the system satisfactory. However, an 11-year-old boy, who had been watching behind me, proved otherwise. After he asked 'Can I have a go?' I gave him the card, but not the code for the keyboard. Nevertheless, he proceeded to open the door. When I asked him how he did it he told me that he had watched which keys I had pressed and he just did the same.

Granted that he was a bright lad, I would still question the value of adding conventional keypads to card-reader systems. It becomes quite a different story if the keys are concealed with a screen surround as shown in *Figure 25.2*.

Plastic cards

These can be encoded in various ways with a personal identity. The use of plastic cards to withdraw cash at bank premises is commonplace. Cards suffer from the same limitation as keys in that they can come into the

Figure 25.2 The risk of unauthorized observation of keyboard entries can be reduced by using a horizontal keyboard with screen surround, as in the Digitac equipment. (Supplied by Raytel Security Systems Ltd)

possession of unauthorised holders and, given no other safeguard, they could be used to obtain the same services as the rightful holder.

Combined keyboard and plastic card selection

In the bank cash-dispenser, the safeguard is usually to combine the plastic card with the keyboard system, thus making it less likely that an impostor could both hold the card and also know the corresponding code to be entered on the keyboard.

If the transferability of plastic cards has to be accepted as a risk corresponding to that with metallic keys, the plastic card is superior in that it can be less easy to duplicate. The relative difficulty of duplication varies with the method of encoding used on the original, and commercial competition is such that there seems to be an endless variety of methods. The very variety becomes a security protection in itself, but the risk of duplication remains, however slight.

The bank cash-dispenser was used to illustrate one way of improving security by combining two separate methods of identification. The keyboard and the plastic card are widely used together in access control systems to establish identity, and their success emphasises the hazards of relying on only one method of identification.

A development that can restore the fortunes of the plastic card is the use of a programmable chip embedded in the card. The information in the chip can be made very difficult to copy and such is the confidence in the method that it is likely to start being incorporated in Bank cash dispensers about the time this third edition is published.

In a version of a Temic chip added security is provided by automatically changing the accept code each time it is used.

A side light showing that constructive thinking is still going on for conventional cards comes from a comment by Mike Smalley of the Security Facilities Executive to the effect that system complexity can be reduced if each card holds only information on the doors that the cardholder is authorised to use, rather than each door having to hold information on all authorised users.

Finger identification

Many attempts have been made to use the individual characteristics of human hands to aid verification of identity. Probably the best known survivor of these attempts is the finger print – an ink impression of a finger tip on paper, widely used by the police. No other significant application of the system has been established due to slowness of application, perceived messiness of the process and the stigma associated with the criminal implication of having one's finger prints taken. However, if a process is quick, clean and effective, the taking of a finger print can be acceptable.

In the Fujitsu 'Fingerscan' equipment, a finger tip can be scanned and compared with a pre-recorded master print in less than half a second. The

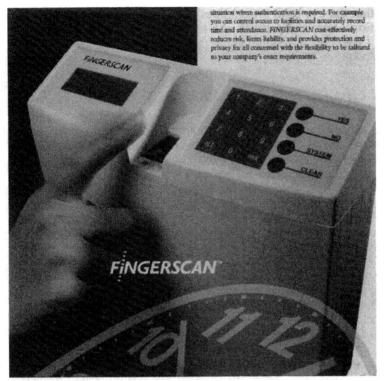

Figure 25.3 Fujitsu Fingerscan

only other time involved in an access control accept–reject situation is for entry of a code on a conventional keyboard to call up the pre-recorded master for comparison. The equipment is shown in *Figure 25.3*.

Voice verification

Until recently, voice verification might have been considered the ultimate we can expect for establishing identity. The method was pioneered with an operational system for the US Air Force Systems Command, and described in various papers by George R. Doddington.

Eye prints

Voice verification may well have been overtaken by events reported in a BBC *Tomorrow's World* TV programme just before I finished writing this 3rd edition. Briefly, it is believed that our eyes are uniquely individual, as are our finger prints. Taking 'eye-prints', both in real time as at an access control point, and for records of authorised personnel, is much more socially acceptable than with finger prints. The BBC demonstration indicated that the eye print system does work. But the second part of the dual question – can it be made not to work – was neither asked nor answered. (See index if in doubt about the 'dual question'.) All being well, eye prints could be in operational use during the life of this 3rd edition.

Error rate

Clearly, a definite mismatch and rejection should occur when an impostor is unable to establish that his keyboard code and his physical characteristics match, and rejection of an impostor is not an error.

Two specific forms of error are possible – either an authorised person can be rejected or an impostor can be accepted. At site entrances it is important to have security personnel on duty to deal with the exceptions and to endeavour to cover the risk of the impostor defeating the automatic selection system. The exceptions include also the authorised people who are rejected by the access control system by mistake, or because they have forgotten their code.

It is customary to express an error rate as a percentage of the total number of people seeking admission. Thus, if 3 out of 100 were rejected, the error rate would be 3%, and at the same error rate 30 people in every 1000 would be rejected. It is easy to see that error rates if used in this way could lead to absurd manning levels or to such disbelief that the system would not be adopted in the first place.

By starting with the user and considering how many security personnel it is reasonable to have monitoring the site entrance at a given time, and then by considering how many people each could sort out if rejected by the access control system, one arrives at the total actual, not percentage, reject rate that can be managed without prejudicing security and without interfering unduly with the overall operation of the site. This information is then fed back to help determine the type of access control that can be accepted.

Single-check systems

Much of what has been said so far regarding access control has been related to double-check systems, such as keypads together with cards. There are many situations where a lower degree of security is acceptable. It may not be wise to set down what we think these situations are but a chief security officer can identify his own. Single-check access control equipment is available for such situations, and a well-known example is shown in *Figure 25.4*.

For establishing identity at site perimeters, card-readers, keyboards and combinations of the two seem to provide the optimum system, particularly as it is so generally accepted that plastic identity cards can legitimately be taken home, away from the site.

For establishments needing restricted access only to certain areas within the site, other methods are feasible, and indeed desirable. Feasibility comes from the fact that, provided that access identification is needed only after being accepted on to the site, tokens that are surrendered each time

Figure 25.4 Wearing a combined identity card/proximity access control token an authorised user approaching a controlled door can proceed freely and open it in the conventional way. (Supplied by Cotag International Ltd)

the holder leaves the site can be used. Given this condition, tokens such as coded pocket radio transmitters can be issued to nominated personnel. These transmitters via their coded identity can then authorise door control systems at specific doorways to allow access to the holder. An authorised person approaching such a door would proceed freely and open the door in the conventional way without the need to re-establish his identity at a card-reader/keyboard post each time.

The number of proximity token systems available is growing apace, and rather than being a security fashion it seems to be on merit. Again, for high-security areas within an outer protected one the risk of the token falling into the wrong hands is minimised in several ways. Given that the token is surrendered to the security personnel at the end of each day before the site is left, re-issue is on a person-to-person daily basis, which reduces the accumulation of opportunities for the token to get into the wrong hands. Second, the number of people authorised to use tokens is likely to be a small proportion of the total number of people on the site, so the prospects of mutual recognition are high and those of noticing a stranger who happens to use a token are correspondingly high.

Discussion points

In dealing with access, and exit, you are concerned with the same people that your fire and safety colleagues are dealing with. Fire and Safety people are endowed with legal powers and obligations which so far are denied to security people. You cannot finalise decisions on your own methods of control without discussing the matter with them. If you haven't done it before, it is important, therefore, to understand the general and detailed principles from their point of view, and a wide-ranging discussion with them is necessary before you come to consider particular cases.

There is a fair amount of documentation available on the subject, and this needs to be studied, particularly if you are at the stage of 'not knowing what you don't know'. National documentation is subject to local interpretation, and local documentation when available needs also to be studied. The combination provides ample material for discussion with fire, safety and security men.

Another point to get clear on is the local view on the access control paradox of automatic control needing manned supervision.

Third, there is the door control problem. Do you know of any situations where it does not match the standard of access control achieved? Discuss with colleagues what each of you would like to see done about it.

Finally, do you see access control as right in concept, and are the equipment suppliers interpreting the need in an acceptable way? What improvements are needed?

26 Personnel and material inspection equipment

International terrorism and the hijacking of aircraft typify risks at one end of the scale, which taper through, according to one's priorities, to such things as theft of materials and tools from a place of work, clothing from department stores and books from libraries.

Physical search

Wherever a person works or visits, the possibility exists that he will see some article that attracts him sufficiently for him to risk taking it away from the premises. Apart from moral aspects, if he is so tempted, a major deterrent is likely to be the prospect of being found out.

Given that management are aware of this type of loss, they may employ security personnel trained in the physical searching of anyone leaving the premises, on a spot check, total check or suspect basis. The likely prospect of a physical search is a definite deterrent, but in situations where the risk is thought to be worth taking people can be almost indecently ingenious in concealing stolen property, to the extent that physical search is not only embarrassing, but also nearly impracticable.

Further, physical searching is expensive in time and manpower, and if ever an accusation is made by a suspect that a security officer has planted an item, it is hard for other people to know what really happened, whatever trust there was is lost and industrial relations between management and personnel become strained. Incidentally, in case of doubt regarding the meaning of 'plant', my dictionary puts it this way: 'a thing, positioned secretly for discovery by another, especially in order to incriminate an innocent person.'

The net result is that physical searching should be adopted only after very careful consideration, particularly on a regular basis, but it cannot be abolished, as nothing else seems to have equivalent detection or deterrent value. So what else is there? Several generic types of electronic search equipment are mentioned below, covering personnel search, baggage, letters and parcels.

Hand friskers

The nearest thing to physical search is the use of hand friskers. One version of this type of device is shown in *Figure 26.1*. In use it needs the

Figure 26.1   A lightweight hand-held metal detector, or 'frisker'. This intimate personnel detector Type IPD/4 by Add-on Electronics is designed to achieve the same objectives as physical bodily search, without exposing the security officer to the risk inherent in physical contact between the officer and a suspect. (Supplied by Adams Electronics Ltd)

full attention of a security officer, who moves the face of the instrument over a person's clothing and body, listening for an audible warning that there is something abnormal about the person being searched. If that person is then asked to explain the alarm or to produce the offending article, the whole search procedure can be carried out without physical contact between the person and the security officer – a wholly more acceptable situation than physical search can give. If the person refuses to cooperate, then a series of actions is open to the security officer which take the situation outside the range of electronic equipment.

When electronic devices are introduced, they have to discriminate between human beings and everything else, if they can. Provided that the object being sought is metallic in nature, electrostatic and electromagnetic methods of detection are available, and they have a reasonably good prospect of working satisfactorily in the vicinity of human beings, who have essentially non-metallic characteristics. If non-metallic materials are being sought, such as drugs, then the authorities may decide that they have no reliable alternative to physical search.

Once, when I was being subjected to a search, I thought I had removed all metallic items from my clothing and had only a finger-ring as metal on my body. But the alarm still sounded. It turned out that what looked like a small plastic stud on my chequebook folder was in fact metal, and the hand frisker being used was well within its detection capability when it picked out the stud.

Arches and pillars

With hand friskers the person stands still and the movement of the instrument is provided by the security officer. Arches and pillars are stationary and the person being searched passes through them, usually with no choice of route if he is to reach his destination. The automatic locking of a turnstile or equivalent will effectively detain a suspect if the alarm triggers as he passes through the instrument.

Arches are general-purpose sensors where the materials at risk are metallic; pillars are more common for detection of non-metallic objects which have metal deliberately inserted in them, such as books in libraries, to make them detectable.

These systems have the advantage of requiring security officer attendance only when an alarm triggers, and therefore one security officer can monitor several arches or pillars simultaneously. These principles having been established, one can accept that more sophisticated versions have been developed for such applications as garment detection in department stores, and the like.

The disadvantage is that usually the distance between the arch or pillar and the metal being sought is greater than exists when the hand frisker is used, and, as we saw in Chapter 4, the inverse square law inevitably tends to reduce the overall detection sensitivity of the system.

X-ray systems for baggage and cargo

The dangers of firearms to pressurised aircraft, and to individual passengers and crew, are such that drastically increased security measures have been necessary at airports for many years. As passengers by the hundred per aircraft and their baggage have to be examined thoroughly yet quickly, almost fully automatic methods are necessary. The X-ray method was a fairly natural choice, and techniques have evolved to a fine art.

Considerable concern was felt regarding the radiation risks involved for operating personnel during direct viewing of the fluorescent screen in their search for hidden weapons. A sizeable step forward was made with the introduction of indirect viewing, with a CCTV camera being used to watch the fluorescent screen and a conventional TV monitor screen in a safe and convenient position.

Although remote monitoring rendered the work safe for operating personnel, concern remained regarding relatively long exposure of the legitimate contents of baggage to radiation, not to mention the risk of damage to any photographic material that happened to be packed away. Quite an elegant solution was found by using microchip memories, rather on the lines used for recording and analysing transient phenomena in the laboratory. In this method the baggage is exposed to X-rays for only a fraction of a second and the resulting information is read off the fluorescent screen TV-wise and stored in a memory. The memory is then read, and displayed for as long as is needed on a TV monitor screen as before. In spite of the repeated transfer of information, the method gives adequately detailed information for its purpose. And to the question from photographers, 'But what about my films?' I could hardly have a better answer. When I put the question directly to International Aeradio Ltd, who specialise in this type of equipment and who supplied the photograph for this chapter, they offered to do a specific test for me. They used a fast, 400 ASA, colour negative film which had been exposed on a typical landscape for several identical frames. They then retained one frame as a control and exposed the remaining frames for 5, 10, 20 and 40 passes, respectively, through their baggage X-ray inspection system, using the

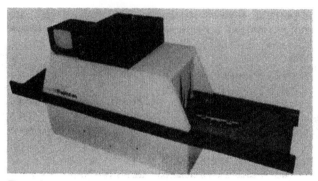

Figure 26.2   The Rapiscan inspects each item in high-definition 'slices' as the baggage passes through a narrow X-ray beam. (Supplied by the Security Systems Division of International Aeradio Ltd)

microchip memory. The prints made from each of the negatives, from the one not exposed to X-rays through to those exposed many times, showed virtually no variation that I could find, other than the minor variations one would expect between several prints taken in succession from the same negative.

Provided that your airport uses modern equipment of this type, your film should be safe. The equipment might look something like *Figure 26.2*.

Yet further development in anti-terrorist-type baggage inspection has led to a three-dimensional viewing X-ray system. With this equipment the user can switch readily from two- to three-dimensional perspective views of baggage contents to help establish their innocence or otherwise. The equipment looks very much the same as that in *Figure 26.2*, and is designed, manufactured and sold by Astrophysics Research Corporation in California and by Astrophysics Research Ltd in Windsor, UK.

Letters and parcels

From the systems described above, all the ingredients are available for checking for letter bombs and parcels containing unwanted items. If metal is a likely constituent of the contents, then a table-top version of the hand frisker is available. For more thorough, and more costly, examination a miniature direct-viewing version of the X-ray baggage inspection equipment is available.

Explosives

One of the non-metals for which a method of detection has been found is explosives. All that is required will go into an executive-type briefcase, and can be taken easily anywhere that explosives may be concealed.

A characteristic vapour surrounds many types of explosive, and with detection sensitivities of better than one part in several million they can be 'sniffed out' by taking a sample of adjacent air and analysing it locally. As with all detection devices, the possibility of false alarms exists, but the consequences are minimised for the equipment covered in this chapter by

Figure 26.3   Typical Cotag tokens. (Supplied by Cotag International Ltd)

the constant presence of security personnel to filter the false from the true. Fortunately, too, for explosive detection, the discrimination is such that it rejects most known domestic and cosmetic vapours.

Other proprietary equipment

Some of the methods developed for hands-free or proximity access control of personnel have been adapted for material identification and material control, and vice versa. Typical perhaps of the 'active' type of token or tag is the aptly named Cotag range. *Figure 26.3* shows two versions of the tag, which need be no larger than a credit card. Each tag is identity coded and is 'read' at critical entry/exit points by wall-mounted sensors. Unauthorised removal of a tagged item can raise an alarm.

Such are the losses from construction sites that the use of active tagging was expanded substantially by Tarmac Construction, through its subsidiary, Castle Site Services. Detection of movement is sensed by selection of equipment from the many devices already available from the security industry – but the movement information is passed via a radio transmitter housed in a compact package, which is fixed temporarily to the machine, tools, materials or whatever is at risk on construction sites, particularly outdoors. Any alarm signal from the transmitter is received by a local monitoring station and passed to the appropriate security forces.

The 'passive' type of token is typified by the Securitag range, developed for theft control in retail premises. For instance, computer tape reels and cartridges can be tagged, and unauthorised attempts at removal of tagged items can be sensed by wall-mounted units or units built arch-like into door frames. These are products of Security Tag Systems Inc. of Florida, distributed in the UK by Security (UK) Ltd, a subsidiary of Automated Security (Holdings) plc.

Discussion points

What other antisocial things and materials need to be detected? How would you use animals, men or electronics to detect them? Is there a point at which the electronic tagging of people becomes unethical?

27 Power supplies

It is perhaps understandable that after a design engineer has worked away at getting an intruder detection system or whatever working satisfactorily, he is happy to leave the design of the power supply for the system to someone else. If it is also thought that there is not much to power supply design, the work may be done by someone who is not adequately trained or experienced or by someone who is too anxious to make cost savings to compensate for some overspending on other parts of the system.

The job may even be put out to a specialist power supply designer and manufacturer. It is possible they know too much about the subject, and the opposite effect of overdesign for the purpose may result.

Power supplies for security systems

As so often is the case, the best way out of this tangle is to find out for oneself something about power supplies. What better place to start than with long-term users of power supplies? With a little prompting they will give you some useful pointers.

User's needs

A few not so fanciful remarks may serve as curtain-raisers to the problem. 'The system has been working perfectly but when we had our first power mains failure in two and a half years last week, the thing went off into alarm almost immediately', or 'The service engineer who is supposed to top up the batteries hasn't been round here for over six months', or 'The fuse went in the power supply unit but we didn't know until the alarm went off because the batteries were flat', or 'Tell me, how can the alarm go off if the batteries are flat?'

Almost everything that follows could go under the heading of 'User's Needs', but we will break it down further to make it easier to follow.

Continuity of supply

All but the mechanical door-bell type of alert needs power in the form of electricity to detect and to signal. And given a security system, what the user needs is that, come what may, the supply of electricity will not fail.

Supplies for signalling

The risk of tampering with anything the villain thinks can get in his way is so real that it is virtually mandatory that all security systems be connected in such a way that if the connections or wiring are interfered with, the signalling system will raise a full alert. The power supply to the system could fail, or be tampered with, yet the alarm must still work. This can most simply be achieved by taking the power supply away from the system and incorporating it in the final signalling equipment (e.g. in a local bell-box mounted in an inaccessible position on the outside wall of a building) and using the same power supply to energise both the bell and the detection system within the building. For larger systems, even if a local power supply is used to energise the detection and local control circuits, the principle of the SAB or self-activating bell has to be followed. That is, that wherever the final alarm signalling is, there also must be a separate power supply unit to energise it. In this way the reaction forces can be alerted even when the supply to the whole of the remainder of the security system has failed.

Sources of power

After many years of negligible development of sources of electrical energy, the needs of outer space equipment and the need for alternatives to oil have spurred on new thinking and the application of known technologies to produce new sources and improvements upon existing ones. Of the four principal sources for security equipment, one is relatively new and the other three have enjoyed significant though invisible improvements.

Primary cells

The characteristics of primary cells are that they are 'dry' and relatively light in weight; consequently, they are widely used for both portable and stationary electronic equipment. From the original wet batteries using carbon and zinc electrodes in ammonium chloride electrolyte (Leclanché) cells, the primary cell has seen a number of evolutionary developments. The Mallory/Duracell manganese type has a higher storage capacity for a given weight and size; the output voltage stays near its rated value for longer than with the Leclanché cell; and, of particular importance in electronic equipment, it is much less likely to discharge corrosive electrolyte through its outer case.

For electronic equipment consuming only small amounts of energy, such as is found in hearing aids, cameras, watches, radio links for intruder detectors and a range of other equipment increasing as energy requirements are reduced, the zinc–air dry primary battery is a further valuable development. Where exceptionally long shelf life is required, or where a high cell voltage combined with a high ampere-hour capacity are needed, as in portable equipment, lithium manganese dioxide batteries provide a versatile alternative.

The one characteristic standing against all primary cells is that once their chemical constituents have released their energy in the form of electricity, they cannot be recharged and have to be replaced.

Solar cells

Although we have to give the Sun pride of place as a source of power, its energy is not available always in the form we want – in this case in the form of electricity, nor at the time we want it. The solar cell as we know it is a direct product of the space age, developed with the help of modern technology into a viable primary source for Earth-based equipment. All the cell needs is light, solar or artificial, to illuminate its face; it then generates electricity that can be extracted from its terminals, as from a chemical primary cell or battery. The word 'battery' means simply a combination of cells connected to give the required voltage and current capacity. Apart from the risk of corrosion, there is not much to stop it going on generating electricity, and lives of tens of years are likely.

Systems and installation design engineers are finding ready uses for the solar cell in security systems indoors and outdoors. It is particularly suitable for situations where it is not practicable or inconvenient or is too costly to provide other forms of supply. As costs come down, it becomes a natural first choice.

The problem here, of course, is what happens at night time, or when there is no light to operate the cell? The next section poses a similar problem, and the answer is covered on page 216.

Mains electricity supply

The first two sources of electricity that we have discussed – primary cells and solar cells – are individual to the user and are capable of supplying only modest amounts of energy, albeit enough for security. For larger amounts of power for lighting, heating and machinery domestically and industrially, it becomes an economic necessity to generate electricity centrally and to distribute it by cable to individual consumers.

You know from Ohm's law that for most conductors such as copper wire the voltage drop E along a length of the conductor is proportional to the current I through it, and the current is inversely proportional to the resistance R of the conductor. Putting these properties together, we have the well-known equation $I = E/R$, or $E = IR$. Power consumption is expressed in watts (voltage E multiplied by current I), so, multiplying both sides of our equation by I, we have $EI = I^2R = W$. We can use this equation to find out how and why power distribution is done as it is.

In power distribution we have not only the power in watts (W in the equation above) consumed by the customer, but also the wasted power lost in the cable from the central power generating station. If R is the resistance of that cable, then from the power equation we can see that if we halved the current in the cable, the power lost in the cable would be reduced to one-quarter. If we did that, we would have to increase the voltage applied to the cable to restore the wattage available to the consumer, and why not? The reduction of waste in distribution cables is so dramatic that for nationwide distribution voltages of 440 000 V and more are used. Power stations generate electricity as alternating current and leave it that way because it is so easy to give consumers the voltage they need by using local distribution transformers. The voltage typically distributed in the UK for lighting and heating, and available for security, is 230–240 V a.c. That, as you will realise, is not the whole story, and you may care to get the gaps filled in during discussion exercises.

The only trouble is that the mains supply sometimes fails due to equipment breakdown, human error or adverse weather, particularly affecting overhead distribution cables and towers. The failure has to be treated as an alert condition, because it may have been caused by an intruder in a forlorn attempt to prevent the system from detecting him. If there was no intruder, the mains failure would have to be treated as a false alarm.

So mains electricity supplies and solar cells have similar problems, and they can share the same solution: the use of secondary cells to tide them over their off-periods.

Secondary cells

The secondary cells best known to the public are car batteries, used for engine starting. Their main virtue is that not only can they store electricity, but also, when the electricity stored has been diminished, the battery can be recharged to full capacity, again and again. Several varieties of secondary cell are available but the type used in cars, although sometimes used for security systems, is not really suitable for that purpose.

Car batteries are designed to power a very heavy load for a short period. Security batteries are needed to supply a light load for a long period. Car owners seem to tolerate a short battery life. Security system owners expect the batteries to last for many years. Car batteries use a wet electrolyte of sulphuric acid diluted with water. In order to recharge them fully ready for the next time they are needed for engine starting, they are charged at a voltage that causes 'gassing' – electrolytic disintegration of the water into its constituents of oxygen and hydrogen. As this process continues, the acid concentration increases and the quantity of electrolyte decreases to the extent that eventually the battery can no longer store electricity – it cannot hold its charge.

In security, although it is often done in error, there is no need to 'overcharge' the battery in this way: we do not need peak voltage.

As we go into our subject in more detail, we find that the differences in requirements do multiply and it is not surprising that design requirements for security are not always understood.

Secondary cells for security

The purpose of secondary cells and batteries in security is to maintain the continuity of the electricity supply to the security system when primary sources such as mains electricity or such alternatives as solar cells are temporarily unable to supply the system.

Solar cells generate electricity already in the form of d.c. needed for operating electronic equipment and charging secondary cells. Mains electricity not only has to be transformed down from the high distribution voltage, but also has to be converted from a.c. to d.c., in rectifier units, as mentioned in the next section.

Although alternatives are coming along slowly, the main choice of secondary cell lies between the nickel–cadmium alkaline cell, the nickel metal hydride cell, each giving a nominal 1.2 V d.c. and the lead–acid cell giving a nominal 2.0 V d.c.

Nickel–cadmium

Ten cells in series are required in order to give 12 V and as they are relatively more expensive than the lead–acid type, this has to be a consideration when making a choice. The advantage to security is that nickel–cadmium batteries are almost non-gassing, which reduces the maintenance attention needed and avoids the derating of cells needed to reduce gassing in lead–acid batteries. If one can take a total cost view over five years or more, nickel–cadmium can show a cost advantage.

Technically it is rather harder to use the advantages of the nickel–cadmium cell. It has very low internal resistance, so it is capable of giving very high output current for relatively short periods. As we saw when looking at the car battery, in security we usually need low currents for long periods. A snag with the nickel–cadmium cell can be that with its low resistance accidental short-circuiting of the battery can be quite spectacular. Also, one has to be careful regarding charging current – without specific current limiting it can easily overload the charging equipment, and if the limiter fails, charger equipment fuses should blow and without further precautions, to be described later, the security system would in due course go into an alarm condition because of lack of power. To be fair, the lead–acid system can suffer in the same way, but for different reasons.

Two more advantages are important. If the battery is outdoors or in a building where the temperature can fall below 0°C, nickel–cadmium cell performance is unaffected. And if the premises have their own stand-by mains generator, there is virtually no reliable alternative to using nickel cadmium batteries for engine starting.

Regarding nickel–metal hydride cells the deletion of cadmium gives an environmental rather than a technical advantage.

Lead–acid

Six cells in series are required in order to give 12 V and with the vast production capability that has been built up to meet the demand for car starter batteries, the price is relatively low. Because people are so familiar with starter batteries, they tend to be unaware that security batteries are used differently and need different treatment.

Open-type lead–acid Perhaps the biggest problem is the inherent property of lead–acid car-type batteries of gassing – releasing oxygen and hydrogen by electrolysis of the water used for diluting the sulphuric acid forming the electrolyte. The lost water has to be replaced by distilled water on a regular maintenance basis, but because electronic equipment needs less attention than mechanical equipment such as cars, topping up is frequently overlooked or neglected. The consequence is that where the battery is suddenly required to take over the supply of power to the security system because of a mains failure, it is in no condition to do so, and it fails shortly after taking over, which leads to a false alarm.

Gassing in open-type lead–acid batteries cannot be overcome except by reducing the voltage at which they are charged. If a battery in a car is charged at up to 2.6 V per cell to give a full charge, a reduction to between 2.35 and 2.4 V is normally recommended for batteries used for security systems. Although this gives a significant reduction of gassing, topping-up maintenance may still be required at least at quarterly intervals. I have found that the only way of getting a reliable extension to topping-up periods is to reduce the charging voltage still further, to 2.2 V.

The penalty for working at a reduced voltage is that the battery can never be fully charged, and therefore its storage capacity or rating in ampere-hours of output is reduced. This can be compensated by using a battery of larger ampere-hour capacity at relatively low cost compared with the cost of servicing and the risk of failure.

Gelled acid batteries Gelled acid batteries are still open and vented, but the acid is absorbed in a jelly to make it less easy to spill accidentally, with consequent risks of corrosion of neighbouring metalwork and equipment. The process does nothing for the chemistry of the battery, and as the 'acid level' is no longer visible, this type of battery seems even more prone to drying out and failure than the open wet battery.

Recombination lead–acid batteries The outstanding development in lead–acid batteries for use in security-type applications is the recombination battery. Only a very small quantity of electrolyte is absorbed into a strip of fibreglass sponge, which is rolled between two electrodes and then sealed into a container. The chemistry is such that, as the gases are evolved, they are retained in the correct proportions and they recombine to form water again. The process is continuous, and the loss of electrolyte through gassing to the open atmosphere is eliminated.

All the virtues of lead–acid batteries are retained, together with longer life under continuous charge conditions; and because there is no gassing, the battery can be operated much nearer to its maximum charge voltage and nearly the maximum rated ampere-hour capacity can be used.

It is important to limit the charging voltage, as excess voltage can lead to overheating of the battery and the simple precautions needed are mentioned in the next section. If a vent is fitted to recombination batteries, it is as a safety precaution in the possible event of overheating and not as a vent for the form of gassing discussed above.

Although nickel–cadmium batteries can offer at least twice the life of sealed recombination lead–acid batteries, they are more expensive; and where first cost is significant in a competitive world, sealed lead–acid seem increasingly likely to be the first choice for the majority of systems requiring reliable support while primary sources such as mains electricity are temporarily unable to supply the system.

For further information try the pioneers of this type of battery – Gates Energy ('Cyclon'), who have manufacturing facilities in the USA and in the UK. F.W.O. Bauch ('Dryfit') make their own version in Germany.

Battery-charging equipment

If mains electricity is to be used, together with secondary cells, then the third element in the power supply system is the battery-charging equipment. The basic elements are a transformer to reduce the mains voltage to near that required by the security equipment and a rectifier to convert the low-voltage a.c. to d.c. To these are added control, protection and indicating equipment.

Mains supply to the charging equipment

Very often, and wisely, alarm installation companies restrict their activities to low-voltage work as needed for their system, and call upon the customer to arrange provision of the mains power supply to the battery charger or rectifier equipment. In specifying to the customer what is needed, two things, in particular, need attention, bearing in mind that some people are interested in causing the supply to fail.

First, the supply should be tapped off a 'maintained' supply. As a fire and safety measure, some establishments cut off all mains electricity supplies to the building during non-working hours. All, that is, except to a few essential services: and arrangements need to be made to regard security as a similar essential service, and the supply should be taken from the 'maintained' supply to avoid the risk of being cut off locally.

Secondly, the villain may be more interested in the rectifier locality, and here the mains should feed straight into the charging equipment housing or, if regulations so demand, via an adjacent 'fused spur' box. The rectifier should not be fed via a wall fixing plug and socket, which can be used so easily to disconnect the supply.

For electronics the power load is likely to be no more than the customer is used to for the simplest of lighting circuits.

The rectifier system

Here the ball is in the electronic designer's court, but he may overlook the need for the a.c. supply to the rectifier to be fused, to avoid consequential damage to a relatively expensive and, in an emergency, difficult to replace transformer caused by a fault in inexpensive components on the load side.

Control equipment

My own feeling is that unless control facilities can be made completely automatic in a very simple way, they should be omitted. The chances of the man who understands manual controls being available when he is needed are relatively low, so any controls provided will be operated to see what happens or not touched at all.

With nickel–cadmium batteries the essential is to have automatic current limiting, to give constant current charging, and with lead–acid batteries the essential is to have automatic voltage limiting, to give constant voltage charging.

To build up a mental picture of what is happening it is best first to consider the system with the battery removed, to emphasise that all the load of the security system is supplied by the mains electricity via the rectifier, without aid from the battery. On 'replacing' the battery, we can see that it 'floats' between the rectifier and the security system, and the voltage across the battery is just sufficient to keep it continuously at the predetermined state of charge. Float-charging in this way leaves the battery in total readiness to take over the supply to the security system in the event of mains failure, without the need for manual or automatic switching.

Indication

On a manned site action can be taken on a fault almost immediately it happens, but on a lock-up-and-leave site there are long periods during which a fault may remain unattended to. In both cases the essential is that those concerned be aware of the fault, and appropriate audible and visual warning indicators are necessary. Firstly, if the battery ceases to charge, because of either mains failure or rectifier system failure, the security system will continue to operate on the battery, but early attention is required. Secondly, prompt attention is needed if the current regulator on nickel–cadmium battery chargers or the voltage regulator on lead–acid battery chargers fails. Ideally, automatic shut-down of the charger should be provided to avoid damage to or destruction of the battery.

Standby duration

The ampere-hour rating of a battery can be somewhat arbitrary, depending upon industry conventions and national standards. Under float-charge conditions it is not practicable to use the whole rated capacity of a battery, so some derating is necessary in determining the suitability of any particular size of battery in security systems. If reliability is expected for several years, a derating factor is needed also for ageing, and a basic figure is best arrived at in discussion of operating conditions with the battery manufacturer. Published data are rarely a sufficient guide.

For 'worst case' design it is possible to obtain from electricity supply authorities the longest duration of supply failure in a particular district over, say, the last 20 years. To guard against a repetition of that failure, the battery can be selected from the derated ampere-hour figure arrived at with the manufacturer.

The situation is complicated by industrial action by electricity supply workers. A typical situation might be 3 hours of supply followed by 3 hours of interrupted supply, and so on. What the future pattern may be depends upon the inconvenience and disruption deemed necessary at the time by the workers, and beyond a certain point designers can do no more than advocate the provision of private local generating plant.

Earthing

Earthing of electronic equipment tends to be neglected in some installations. Experience indicates that a range of elusive types of false alarm can be eliminated by earthing the d.c. side of a security system power supply. It is traditional to earth the negative, but provided that it is compatible with the electronics used, there is evidence that earthing the positive produces less corrosion. So long as the d.c. is earthed, the choice lies perhaps with local convention.

Direct current earthing is not to be confused with what is normally mandatory earthing of mains-connected equipment, and it can be beneficial to keep the two earth connections separated. Similarly, where the mains operated equipment is 'dual insulated' and not earthed the d.c. supply to the security system should, nevertheless, be earthed.

Power supplies for door control

Electronic equipment for security in general consumes only low values of current, typically at 12 V d.c., and there rarely are problems of excess voltage drop in cabling to equipment over quite large areas of buildings. Exceptionally, there can be problems in providing sufficient power economically for operating solenoids for door control in access control systems. Where a power supply has to be some distance away from the door or doors to be controlled, it is preferable to minimise cable voltage drop by increasing the distribution voltage.

Most authorities accept up to 50 V as 'low-voltage' distribution, and some will accept an earthed centre-tap on a transformer giving up to 110 V. Even an increase to 24 V, as used widely for fire alarms, produces a significant saving in voltage drop, and once the principle is accepted of changing from a standard voltage for such things as solenoids, there is rarely a problem in deciding upon the right figure to operate at.

Protection

Rounding off the theme that anything extra that is provided in power supplies for security equipment should operate automatically, protection is needed for semiconductor equipment against severe overvoltage spikes that occur on mains electricity supplies. Protection can be given by using 'Zener-type' devices, which present a high resistance to normal voltages, but a much reduced resistance to predetermined excess voltage, thus effectively reducing the effective amplitude of the spike. Overvoltage surges last longer than spikes, usually with a lower peak amplitude, so in selection of protection devices consideration has to be given to the total energy to be diverted. Specialist design of Zener barriers can also be used between power supply units and areas requiring intrinsic safety.

Electromagnetic compatibility (EMC)

In 1996 European regulations came into force defining limits of interference caused by electrical/electronic equipment and also defining acceptable limits of immunity of equipment to interference. The regulations can affect all equipment used in security, but a key factor affecting applicability is the environment in which the equipment is to be used. Up-to-date guidance should be obtained from an independent body such as the Electrical Research Establishment (ERE), Leatherhead, Surrey, UK.

Ventilation

With unsealed lead–acid batteries potentially dangerous concentrations of hydrogen and oxygen can accumulate in unventilated cupboards and housings for battery-charging equipment. Even 'sealed' units usually have safety vents. Care is needed, therefore, in design and installation to ensure that adequate ventilation is provided for battery supplies for security equipment.

Discussion points

As has been illustrated above, power supplies for security are by no means the simple affairs they, perhaps, should be. The demands upon them differ from those made by many other types of equipment and solutions do not seem to fit an exact science. There is room, therefore, for opinion and choice. You may care to enlarge on the relative merits of lead–acid and nickel–cadmium batteries, and see whether within your own organisation there is a common view, or divided opinion. In either case the situation may not coincide with the existing corporate policy. If so, your discussion might go on to assess whether you can, or even whether you should, try to have that policy changed to match your opinion as you see it today.

28 Standards

'Who needs standards?' is the theme of a document from the British Standards Institution. They answer their own question by saying 'You do!'

The aim of the document is to guide businesses and others in developing their own standards policy. To get your copy, apply to:

The Information Officer
British Standards Institution
389 Chiswick High Road
London W4 4AL

As you read it, remember though that standards are 'A guide for the time being'. How else could new ideas develop into products and services, and, of course, into new and revised versions of standards and codes of practice? While the standards are current and applicable, however, they can have the force of law and have to be obeyed.

When I was helping in the drafting of the original standard for 'Intruder alarm systems in buildings' BS 4737, little did I realise the vast array of security standards that were to follow. As in so many avenues of thought, the reasons for, and an understanding of, the need for implementation of standards are learned by starting with the end-user and then working inwards towards the designer and laboratory.

Security is no exception and in the UK the most specific requirements for the design and use of security equipment have come from the police, who in this case do represent the user. In their exasperation at the continued high rate of false alarms, they have issued, under the authority of the Association of Chief Police Officers, a document entitled 'ACPO Intruder Alarm Policy.' Although not a standard in the normal sense, it has to be regarded now as the foundation standard from which all else follows.

Many standards, however, did exist long before the police became so intimately involved. A fairly comprehensive listing follows, together with a flow chart, which helps in checking that you have selected the most appropriate references for the job you have in hand at the time. The listing has been grouped in the main to give the subject first, and then the appropriate number for tracking it down in a library etc. and for purchasing.

Almost all of the data listed have been contributed by Mike Cahalane, Chief Executive of Mica Associates, Security Consultants of London W5, who very kindly offered it for this chapter on security standards. USA and world readers will I hope forgive the omission of their standards. Their inclusion would necessitate a new Part 4 to the book. I would prefer that this comment were a cue for someone to write a whole book devoted to security standards.

Standards and codes of practice

Categories listed

		Standards	Codes of practice
A	Electronic	Standards	Codes of practice
B	Electrical	"	"
C	Mechanical	"	"
D	Quality control	"	—
E	General	"	"
F	EMC	"	—
G	BSIA	"	"
H	European	"	—

A. Electronic – Standards

1.	Alarm terminating equipment in police stations	BS4166
2.	Printed circuits, base material for metal cladding	BS4584
3.	Security systems with local and remote signalling	BS4737 Pt.1
4.	Radiowave doppler detectors	BS4737.3.4
5.	Ultrasonic movement detectors	BS4737.3.5
6.	Acoustic detectors	BS4737.3.6
7.	Passive infrared detectors	BS4737.3.7
8.	Volumetric capacitive detectors	BS4737.3.8
9.	Beam interruption detectors	BS4737.3.12
10.	Capacitive proximity detectors	BS4737.3.13
11.	Specification for sound level meters	BS5969
12.	Requirements for connection to telecommunication networks	BS6301
13.	Requirements for connection to British Telecommunications telephone network	BS6305
14.	Modems for connection to telecommunication network	BS6320
15.	Intruder alarm systems for consumer installations	BS6707
16.	Home and personal security devices	BS6800
17.	High-security alarm systems in buildings	BS7042
18.	Sound systems for emergency purposes	BS7443
19.	Specification for multi-frequency tone signalling protocol for social alarm systems	BS7369

A. Electronic – Codes of practice

1. Wire-free alarms	NACP12
2. Telephone line fault monitoring	NAD1
3. Model paragraphs regarding movement detectors	NATM3
4. High-security premises	SCOP105
5. Wire-free interconnections within intruder alarms	SCOP106
6. Alarm system, planning and installing	BS4737.4.1
7. Alarm system, maintenance and records	BS4737.4.2
8. Exterior alarms	BS4737.4.3
9. Remote centres for alarm systems	BS5979
10. Planning and installation of sound systems	BS6259
11. Wire-free intruder alarm systems	BS6799
12. Social alarm systems	BS6804
13. Integration of fire and security systems	BS7807
14. Article theft detection systems	BS7230
15. Exterior deterrent alarms (see BS4737.4.3)	DD231
16. Management of subcontracting	NACP3.2
17. Management of customer complaints	NACP5
18. Management of false alarms. Plus AMD1	NACP10
19. Intruder alarms for high-security premises	NACP13
20. Intruder alarm systems signalling to alarm receiving centres	NACP14
21. Time delays in remote signalling by digital communicator	MATM1

B. Electrical – Standards

1. Protection against electric shock	BS2754
2. Safety requirements for electrical appliances	BS3456
3. Isolating transformer safety requirements	BS3535
4. Continuous wiring	BS4737.3.1
5. Protective switches	BS4737.3.3
6. Pressure mats	BS4737.3.9
7. Vibration detectors	BS4737.3.10
8. Specification for PVC cables	BS4737.3.30
9. PVC cable requirements for telecommunications	BS4808.1
10. Arc welding of carbon and manganese steels	BS5135
11. Conductors in insulated conductors and cables	BS6360
12. Safety requirements for power supply units (PSU) for equipment connected to telecommunication networks	BS6484
13. PVC insulation and sheath of alarm system cables	BS6746
14. Requirements for apparatus to be connected to telephone networks	BS6789
15. Connection of social alarms to public telephone networks	BS7606
16. Electrical installation (was IEE Wiring Regulations)	BS7671
17. Electrical actuation of gaseous total flooding extinguishing systems	BS7273.1

B. Electrical – Codes of practice

1. Emergency lighting of buildings   BS5266
2. Electrical equipment for use in explosive atmospheres BS5345
3. Safe operation of alkaline secondary batteries   BS6132
4. Safe operation of lead–acid secondary batteries   BS6133
5. Intruder alarm systems with mains wiring
   communication   BS7150

C. Mechanical – Standards

1. Glass for glazing   BS952
2. Expanded metal   BS405
3. Timber for joinery   BS1186
4. Wood screws   BS1210
5. Metal door frames   BS1245
6. Aluminium for rivets, bolts and screws   BS1473
7. Chain link, bar and palisade fences   BS1722
8. Environmental testing: basic procedures   BS2011
9. Thief-resistant locks   BS3621
10. Dimensions for resawn softwood   BS4471
11. Foil on glass   BS4737.3.2
12. Wood doors and frames, internal and external   BS4787
13. Aluminium alloy windows specification   BS4873
14. Bullet resistant glazing for interior use   BS5051.1
15. Bullet resistant glazing for exterior use   BS5051.2
16. Degrees of protection provided by enclosures   BS5490
17. Anti-bandit glazing   BS5544
18. Locks and latches for doors in buildings   BS5872
19. Impact performance of flat safety glass and plastics BS6206
20. Steel windows, sills, window boards, doors   BS6510
21. Security seals   BS7480
22. Guide for security of buildings against crime   BS8220

C. Mechanical – Codes of practice

1. Installation of security glazing   BS5357
2. Glazing for buildings   BS6262

D. Quality control – Standards

1. Guidelines for developing quality manuals   BS ISO 10013
2. Quality specifications for design (was BS5750.1)   BS.EN.ISO.9001
3. Quality specifications for manufacture and
   installation (BS5750.2)   BS.EN.ISO.9002
4. Quality specifications for final inspection and test BS5750.3
                                                                              BS.EN.ISO.9003

5. Guide to the revision of BS5750/ISO.9000	BS5750/ISO.9000
6. Reliability of systems, equipment and components	BS5760.1
7. Quality management – system elements	EN.ISO.9004.1
8. Generic guidelines for application of ISO.9001/2/3	DIN.ISO.9000.2
9. International environmental management	ISO.14001
10. Glossary of terms used in quality assurance	BS4778
11. Financial transaction cards. Security architecture of financial transaction systems using integrated circuit cards	BS.EN30202
12. Guide to achieving compliance with EC directives for alarm systems PD6608	CENELEC report RO79.001
13. Quality schedule for the installation and maintenance of security systems issue 2	SSQS.101
14. Quality schedule for central stations to BS5979	SSQS.102
15. General criteria for operation of testing laboratories	EN45001
16. General criteria for assessment of testing laboratories	EN45002
17. General requirements for testing laboratory accreditation	EN45003
18. General criteria for operation of inspection bodies	EN45004
19. General criteria for suppliers declaration of conformity	EN45014

Note: On quality and related standards, the joint European Standards Institution CEN/CENELEC is currently re-issuing some existing standards with new numbers. Refer to BSIA or NACOSS if confused.

E. *General – Standards*

1. Guide for graphical symbols for electrical power, telecommunications and electronic diagrams	BS3939
2. Glossary of terms for electronics, telecommunications and power	BS4727
3. Specification for installed systems for deliberate operation	BS4737.2
4. General requirements	BS4737.3
5. Deliberately operated devices	BS4737.3.14
6. Recommendations for symbols for diagrams	BS4737.5.2
7. EMC guidelines for installers	NATM6

E. *General – Codes of practice*

1. Intruder alarms for high-security premises	NACP13
2. Management of false alarms	NACP10

3. Noise and Statutory Nuisance Act 1993   Schedule 3
   Noise from Intruder Alarms   Section 9
4. Control of Pollution Act 1964 Alarms   Time Limit
5. Keyholder training and responsibilities   ABI Guidance
6. Security screening   NACP1
7. Customer communications   NACP2
8. Management of sub-contracting   NACP3
9. Compilation of control manual   NACP4
10. Planning installation and maintenance of
    intruder alarms   NACP11
11. Planning installation and maintenance of CCTV
    systems   NACP20
12. Manning, installation and maintenance of access
    control systems   NACP30
13. Keyholders. Central station policy   NAD2
14. Disconnection of alarm monitoring services   NATM5
15. Performance of alarm installers and maintainers   NATM7
16. Management of false alarms   SCOP104
17. Manned security systems   BS7499
18. Documentation for social alarm systems   BS7524
19. Information. Security management   BS7799
20. Security screening of personnel   BS7858
21. Installation of apparatus intended for connection
    to certain telecommunication systems.   BS6701
    Specification for general requirements   Part 1
22. Installation of switching apparatus
    Apparatus that may be connected to certain   BS6701
    analogue telecommunication systems   Part 2
23. Rules relating to issue of NACOSS compliance
    certificates   NAD3

F. Electromagnetic compatibility (EMC)

In 1996 European regulations came into force defining limits of interference caused by electrical/electronic equipment, and defining also
acceptable limits of immunity of equipment to interference. The regulations can affect all equipment used in security, but a key factor affecting applicability is the environment in which the equipment is to be used.
Up-to-date guidance should be obtained from an independent body such
as the Electrical Research Establishment, Leatherhead, Surrey, UK. List
abstracted from the Official Journal (97.C270/06) and published by
Maud M Business Communications, Sevenoaks, TN13 2YH, UK, in
'approval', an engineering guide to European quality standards and
regulations.

1. Signalling on low voltage electrical installations
    in frequency range 3–148.5 kHz   EN50065-1
2. Generic emission standard: residential,
    commercial and light industrial environments   EN50081-1

3. Generic emission standard: industrial
   environments   EN50081-2
4. Generic immunity standard: residential,
   commercial and light industrial environments   EN50082-1
5. Generic immunity standard: industrial
   environments   EN50082-2
6. Cable distribution systems for television and
   sound signals   EN50083-2
7. Home and building electronic systems (HBES)   EN50090-2-2
8. Uninterruptible power systems (UPS)   EN50091-2
9. Immunity for components of fire, intruder and
   social alarm systems   EN50130-4
10. Electronic taximeters   EN50148
11. EMC product standard for arc welding equipment EN50199
12. Industrial, scientific and medical (ISM) equipment
    (emissions)   EN55011
13. Broadcast receivers and associated equipment   EN55013
14. Household and similar electrical appliances   EN55014
15. Fluorescent lamps and luminaires   EN55015
16. Immunity of broadcast receivers and associated
    equipment   EN55020
17. Information technology equipment emissions   EN55022
18. Audio, video and entertainment lighting control
    emissions   EN55103-1
19. Audio, video and entertainment lighting control
    immunity   EN55103-2
20. Immunity of household and similar appliances   EN55104
21. Low-voltage fuses   EN60269-1
22. High-voltage fuses   EN60282-1
23. Watt-hour meters   EN60521
24. Mains harmonics in household appliances
    (will be superceded by EN61000-3-2)   EN60555-2
25. Flicker in household appliances (will be
    superceded by EN61000-3)   EN60555-3
26. Medical electrical equipment   EN60601-1-2
27. a.c. static watt-hour meters   EN60687
28. Automatic electrical controls for household
    and similar use   EN60730-1
29. Automatic electrical burner control systems   EN60730-2-5
30. Automatic electrical pressure sensing controls   EN60730-2-6
31. Timers and time switches   EN60730-2-7
32. Electrically operated water valves   EN60730-2-8
33. Temperature sensing controls   EN60730-2-9
34. Energy regulators   EN60730-2-11
35. Telecontrol equipment and systems   EN60870-2-1
36. Marine navigation equipment   EN60945
37. Low-voltage switch gear and control gear   EN60947-1
38. Circuit breakers   EN60947-2
39. Switches, disconnectors, switch disconnectors
    and fuse combination units   EN60947-3

40. Electromechanical contactors and motor starters   EN60947-4-1
41. Electromechanical control circuit devices   EN60947-5-1
42. Proximity switches   EN60947-5-2
43. Automatic transfer switching equipment   EN60947-6-1
44. Control and protective switching devices   EN60947-6-2
45. Mains harmonics   EN61000-3-2
46. Mains flicker   EN61000-3-3
47. Residual current circuit breakers   EN61008-1
48. Residual current circuit breakers with integral
    over-current protection   EN61009-1
49. a.c. static watt-hour meters   EN61036
50. Electronic ripple control receivers for tariff and
    load control   EN61037
51. Programmable controllers   EN61131-2
52. Residual current protective devices   EN61543
53. Equipment for general lighting purposes
    (immunity)   EN61547
54. Adjustable speed electrical power drive systems   EN61800-3
55. Susceptibility of industrial process and control
    equipment to electrostatic discharge   BS6667-2
56. Susceptibility of industrial process and control
    equipment to radiated electromagnetic energy   BS6667-3
57. Guidelines for installers regarding EMC   NATM6
58. For guidance booklet on UK EMC compatibility
    telephone (0117) 944 4888
59. Electromagnetic Compatibility Council Directive   89/336/EEC
60. Amending Directive   92/31/EEC
61. European EMC product family immunity standard BSEN50133.1
62. European EMC alarm immunity standard   EN50 130.4

G. BSIA – Standards and code of practice

List contributed by BSIA	BSIA No.
1. Access control systems. Planning, installation and maintenance	107
2. CCTV systems. Planning, installation and maintenance	109
3. False alarm management	110
4. Management of sub-contracting	128
5. Minimum standard for operational control rooms (guard and patrol section only)	136
6. Management of CCTV sub-contracting	140
7. Guide to the selection and installation of movement detectors	147
8. Alarm user guide to the ACPO policy	154
9. Long-range radio systems	161
10. BFPSA/BSIA integrated fire and alarm systems	162
11. Guide to special information tones	165
12. Minimum training standards. Static and patrol guarding	168
13. EMC guidelines for installers of security systems	195

H. European – Standards

Standards published (1997)

12. R79-001 Guide to achieving compliance   PD 6608
13. Secure storage units. Resistance to burglary   BS EN 1143-1

Standards under preparation
1. BS/Euro version of BS4737   EN50 111.3.1

Alarms
2. Introduction to series of standards   EN50 130.1
3. Intrusion. General requirements   EN50 131.1
4. Intrusion. Application guidelines   EN50 131.7
5. Hold-up. Systems requirements   EN50 135.1
6. Hold-up. Trigger devices   EN50 135.2
7. Hold-up. Application guidelines   EN50 135.7

Detectors
8. Common requirements   EN50 131.2.1
9. PIR   EN50 131.2.2
10. Microwave   EN50 131.2.3
11. Combined PIR–microwave   EN50 131.2.4
12. Ultrasonic   EN50 131.2.5
13. Opening contacts   EN50 131.2.6
14. Glass break, acoustic, seismic   EN50 131.2.7
15. Vibration   EN50 131.2.8
16. Active infrared   EN50 131.2.9
17. Proximity   EN50 131.2.10
18. Others   EN50 131.2.11

Alarms
19. Intrusion and hold-up. Control and indicating
     equipment   EN50 131.3
20. Intrusion and hold-up. Power supply equipment   EN50 135.5
21. Requirements for dedicated wired links   EN50 131.5.1

Social
22. System requirements   EN50 135.1.1
23. Trigger devices   EN50 134.2.1
24. Trigger requirements, inactivity   EN50 134.2.2
25. Local unit and controller   EN50 134.3
26. Warnings   EN50 134.4
27. Interconnections and communications   EN50 134.5
28. Power supplies   EN50 134.6

Transmission
29. Annunciation equipment   EN50 136.4
30. Equipment. General requirements   EN50 136.2.1
31. Dedicated alarm paths   EN50 136.2.2
32. Requirements for digcoms   EN50 136.2.3
33. Voice communications   EN50 136.2.4
34. General requirements   EN50 136.3
35. Application guidelines   EN50 136.7

Alarms
36. Intrusion. Warning devices   EN50 131.4
37. Hold-up. Warning devices   EN50 135.4

CCTV
38. System requirements   EN50 132.1
39. Colour cameras   EN50 132.2.2
40. Lenses   EN50 132.2.3
41. Ancillary equipment   EN50 132.2.4
42. Local and main control equipment   EN50 132.3
43. Black and white monitors   EN50 132.4.1
44. Colour cameras   EN50 132.4.2
45. Recording equipment   EN50 132.4.3
46. Hard copy equipment   EN50 132.4.4
47. Video monitor   EN50 132.4.5
48. Video transmission   EN50 132.5

Access
49. General requirements for components   EN50 133.2.1
50. Processing equipment display   EN50 133.3
51. Access point actuator   EN50 133.4
52. Communicator   EN50 133.5
53. Application guideline   EN50 130.7

Alarms
54. EMC tests for immunity   EN50 130.4
55. Environmental tests   EN50 130.5
56. Requirements for dedicated wired links   EN50 131.5.1
57. Requirements for non-dedicated wired links   EN50 131.5.2
58. Requirements for RF links   EN50 131.5.3
59. Combined/integrated systems   TC79/SEC/255

Discussion points

You will realise that this book sets out to encourage new ideas aimed at improving crime prevention and at reducing false alarms. Standards, by their very nature, can work against these objectives. It is important, therefore, for people writing new standards to be aware of the problem, and so find words which achieve the objective of control, yet do not inhibit progress that could be possible through advances, say, in technology. If you were writing a standard on how to write standards, how would you word the paragraph illustrating the dilemma, and how, by skilful wording, the problem can be overcome?

Part 3

Implementation

29 Presentation of information

Some years ago a young student visiting England stayed with our family for a few days. His subject was 'Communications', and I soon learned from him that he meant something quite different from my definition of the word. My degree subject also was Communications – Electrocommunications, dealing with techniques for transmission and reception of signals by telegraph, telephone and radio. Our student friend, on the other hand, was learning techniques for speaking, writing and visual display for communication to people, via the press, the radio and television.

His discipline was an end in itself, mine was a means to an end; and the two, though so different, are complementary. In security we find that we get involved not only with the transmission and reception of alarm signal information, but also in the presentation of information in visual display, and audible and written form.

The purpose

The purpose of presentation is to convey information to selected people in such a way that they can assimilate it readily and act upon the information promptly, confident that they understand it correctly. For an author to write about presentation of information is to invite the comment from the reader, 'I wish he would practise what he preaches'. If I am to take that risk, I must thank Professor R.O. Kapp for making me aware of the need to study the subject in a postgraduate course he set up entitled 'The Presentation of Technical Information'.

The essence of a subject can often be expressed in a few words and for some people that is enough to trigger off their minds to think of all the ways the concept can be applied to their work. Others, with perhaps less vivid imagination or experience, find it helpful to have the subject expanded by description and case histories to bring out the validity and relevance to them in their own lives.

A notable example is value engineering. The essence of that concept is to examine an existing design or product and to find ways of reducing by a given percentage the total cost of its manufacture, without depriving the user of any feature that he valued. This is such a natural idea and can be expressed in so few words that it is hard to understand that it was not standard practice in industry, but it was not. Once expressed, the idea of value engineering caught on in an industrial world fighting to retain its markets. Whole books were written about it, and new businesses were set up to help firms to implement it.

The concept of presentation of information is even more simple, and, for some, once it has been expressed nothing more need be said. For most people, however, it is helpful to expand the concept with illustrations of its meaning. You see, the reason why the concept of value engineering took off was that it was presented in a way that the reader, the user of the information, could adopt as his own: he could relate to it. He could see how it applied to him and he could apply his new understanding in his work, and, above all, he found that it did what it was claimed to do. Thus, confidence in the idea was established.

Now we have all the ingredients of the concept of presentation of information, whether it be purchasing advice to management, selling to potential customers, system design information to the engineering department, manufacturing information for the factory, field engineering standards for the installation department, or operational orders for uniformed security staff.

You will see that in each case I have said who the information is for. That is the starting point, the user. A wise chief security officer once said to me, 'Tell me what you want my men to do, and I will write an order in words that they will understand'. That is good advice. If you have doubts about being able to get your message across, find, if you can, someone who understands you and who understands the user so well that he speaks his language. If you have to do it yourself, then this chapter should help you to make a better job of it. Let us work through a few examples.

Examples

Information to management

Management is your boss and his boss. They are there to get defined results, usually meaning financial success, which sooner or later means selling something. They have people to help them, maybe production people, probably sales people. These are the mainstream needs of a business.

Enlightened management concentrate not only on the direct objective of getting things done, but also on things that interfere with getting things done. If a vital raw material disappears from the stores overnight, factory output will fall below target. A security man may be employed to help guard against this risk to production, but management is unlikely to regard the security man as a part of the mainstream activity. The security man is likely, therefore, to have to accept the role of adviser to management rather than that of a mainstream executive, and to that extent he is regarded as non-productive. To overcome this apparent disadvantage, the security adviser has to show that he can and does think in mainstream terms, as well as showing that he understands his own subject. An example given in Chapter 2 when we considered the need to define the problem may be worth repeating here, with perhaps different emphasis to bring out here its relevance to communication with management.

In the example we found a chief security officer going to his director to report the loss of valuable stock and meeting with an unsympathetic reply that it had happened before and what was the CSO going to do about it? Because the CSO was not prepared for that question, he had succeeded only in aggravating his director, and had failed to get authority to have changes made that would improve security.

Anxious to retrieve the situation, he set about doing what he should have done in the first instance. Fortunately, he was able to establish quite quickly that it was a daytime loss. Concentrating on this one aspect of the problem, the CSO traced through every step in the chain from executive decision to purchase additional stock, through the transport loading bay and goods inwards, to payment of invoice for goods received and claims on insurance for shortages. He found that systematic information on dates, times and quantities of despatch and arrival were being issued to personnel who had 'no need to know' the information.

Armed with these findings, the CSO was able to go to his boss with the advice that if the stock movement information was restricted to those with a definite need to know, he believed the losses would stop. In addition, it would not, after all, be necessary to have CCTV monitoring of the loading bay as originally thought, and the elimination of unnecessary paperwork would reduce costs.

By presenting the situation not as a problem but as a proposed solution which would not only stop the losses, but also reduce operating costs, the CSO was talking a language understood by and welcome to his director. The outcome was ready agreement to have the paperwork routine changed, discreetly, and the losses ceased.

Information to control room personnel

Although, perhaps, at the other end of the scale from the director in the example above, the officer on duty in a security control room is one of the most important 'users' in operational security. In sorting out how to present information to him, it is necessary to understand what operational life means to him.

By its nature it tends to have military undertones, be it in a large company's own control room, or in an alarm company's central station serving many customers, or in a police station serving the local community. It is hardly surprising, then, that operational personnel's standing orders on what to do are expressed in military terms in typed or written form. But, traditionally, when associated with electronic security systems, the information on when and where action is to be taken is given by signal lights and audible alarms. There can be long periods when operational personnel have no alarm to attend to; yet, when an alert is raised, security officers are expected to respond instantly and correctly. Thus, there is scope for ensuring that the information they receive is in a form that they can comprehend accurately and accept with confidence.

For 'in-house' control rooms the idea of mimic panels, also called graphic panels, has been established for many years. Here a wall- or desk-mounted plan picture of the premises is set up adjacent to the duty officer, so orientated that the officer can relate instantly to a light flashing on the panel – in the top left-hand corner, say. Orientation is important, although it tends to be taken for granted. Its importance is emphasised when one feels intuitively that the mimic panel has been designed and mounted 'upside down', and valuable time can be lost in locating the trouble spot in one's mind.

Further aids in presentation of information on a mimic panel include separate colours to distinguish security from fire and from other emergency situations. The argument against this is that not all people have the same mental response to different colours – they may be 'colour blind'. The next step is to add legends to the indicator lights, saying, for instance, 'Boiler House'. If the legends are added generously, they become self-defeating – the mimic panel has to be made larger to accommodate all the information, which has to be there all the time even though only one item of it is required for any given incident, and the stage is reached when normal eyesight cannot cope with reading the legends at various distances from the duty officer's control position.

Electronics now solves the problem in a most attractive and confidence-giving way. If a video display unit or TV screen is used, an incident can cause a prerecorded or prestored plan picture of a risk area to be displayed on the screen. Unlike the situation with a mimic panel, only that part of the area needed for orientation and identification need be shown on the screen, which leaves plenty of room for written information to be displayed giving the location of the incident and its nature – intruders, fire, and so on; and, most important, instructions on the actions the duty officer must take can be displayed visually at a convenient viewing distance.

In addition to these advantages of electronic display of information, all information needed for taking remedial action can be printed out automatically on paper, which can be handed to other personnel for attention as instructed.

Finally, the print-out enables an 'office copy' to be retained giving a permanent record of the time and location of each incident.

Information for manufacture

A sobering thought for electronic designers is that nothing we do has an inherent existence of its own. The only way we can give permanence to electronics is to express it in physical, mechanical form. To put this idea into words appears only to be stating the obvious, but there is ample evidence that its implications are not automatically understood and acted upon. The electronic designer who recognises this recognises that he has a profound vested interest in the presentation of information not only *to*, but also *by*, mechanical designers who are concerned with expressing electronics ideas in physical usable form.

To take an elementary example, how often have you stood or sat in front of some electronic equipment, unable to find the control you want, not because it wasn't there but because the labels identifying the controls were underneath the knobs and were invisible from a normal operating position. Only by ducking your head can you read the labels.

On the drawing-board it is only too natural to locate the identity wording below the control, because everything is in two dimensions and only when the item is created in three dimensions do the problem arise. Why don't they put identity labels above the controls if that presents the information more readily to the user?

The right attitude comes when it is accepted that drawings are not just records of information but are also instructions on how things are to be made.

It is sometimes argued that a good drawing should speak for itself and should contain the minimum of words. However, I have never found it a disadvantage to add notes to drawings to explain why something was required to be done in perhaps an unusual way.

Getting decisions on implementation

Examples such as these at various levels of operation all show the same thing. It is not enough to transmit a message. If it is to be received, understood, accepted and acted upon, the message has to be put across in the right form. Realisation of this fact and liaison with the people concerned are half the battle towards getting the message across successfully.

However, in the first example getting agreement to a course of action that saves money should have been easy anyway – the problem was to get the man to listen. What if you want authority to spend money – on a new security project perhaps?

It is worth remembering that your boss may have to sell your ideas to the financial director and the managing director, so he needs ammunition. Think of him as being on your side, and feed him with facts and information that he can use in his own way. He has probably got where he is by knowing how to gain attention, and he does this by proper preparation. Why not learn from him? As an adviser you are expected to be skilled in the art already, but there is always something new to be learned. Normally one needs some way of relaxing the atmosphere and removing his concentration from other pressing problems. One worthy boss of mine, knowing the dilemma, used to jump the gun and fire 'How's the family?' at me as an opening shot. So be prepared for that too; don't let it deflect you from saying what you came to say, but be sure you have your boss's attention before presenting your case.

The framework, then, is to reiterate the need; remind him of the normal course of events and then introduce the problem – what is interfering with that course and what is needed to improve the situation from the firm's point of view, whether that be changes in method, some capital expenditure or an increase in manpower.

Nearly everything is eventually related to cost, and it is worth knowing how much expenditure one's boss himself can authorise. In security it is quite hard to relate potential savings to actual costs in a convincing manner, and there is no harm in working a little on the possible consequences of not taking your advice. No one is particularly keen on the prospect of being told later, 'I told you that would happen if we didn't take such and such precautions'. Likewise, if you are persuading a customer to spend money, remember that he, too, probably needs ammunition to persuade his boss; think of him as being on your side, selling your ideas.

Discussion points

If your reaction on reading this chapter is that none of the ideas would work on your boss, your staff or your customer, good – you have a creative starting point. If it's not too personal a question – What would work? The one thing that is certain in this context is that security is a communications problem, and as such it involves people.

Whether it be from employer or employee, one so often hears the frustrated comment 'Why don't they' or 'Why do they', because they have not been told, or not told in a way they understand. How do you think information should be presented so that the user can truly say, 'Message received'?

30 System design

'Why ever did they do that?'

I went to see a building a few weeks ago that had suffered a series of break-ins, and my mission was to vet proposals that had been made for provision of an electronic detection system. Before I had even gone into the building, there in front of me was the awful sight of a recently fitted grille security door with the steel facing ripped off and the dead-lock bolt sprung away from the door-frame.

The specification offered by the supplier of the door was shown to me, and it spoke of a welded steel angle-frame, bars welded into the frame of the door itself and the door clad overall in steel sheet. The purchaser could be forgiven for assuming that the steel sheet also was to be welded on. But it wasn't: it was 'fixed' by pop-type rivets, and the intruder had no difficulty in de-popping them, in removing the steel sheet and gaining access to the dead-lock. The second error by the supplier of the door was to provide and fit a lock with an inadequate length of dead bolt protruding through into the stationary frame of the door. Why ever did the supplier overlook these two methods of attack?

Part 3 is concerned with Implementation, and this chapter is concerned with establishing a requirement for system design, the responsibilities of the people who do it, and the responsibility of the supplier to implement the system in a manner that conforms with sound security engineering practice, not to mention the various Codes of Practice and Standards that now apply.

Who can help the customer?

The whole of Part 1 was devoted to Security Systems, and forms a necessary foundation for what follows. There cannot be a system without a customer requirement, and the case history just given shows how difficult it is for the customer to start aright. They had had losses; they had employed someone to deal with the problem; but that someone did not, or did not know how to, deal with the problem adequately. If a person is unaware that dealing with security requires some specialised knowledge and ability, he is isolated from our world and there is little or nothing we can do about it.

How, then, is the potential customer to know who to approach for advice? One way of getting through to people who do not know the extent of their need is by advertising. It used to be said, and it is still largely true, that advertising a particular security service or product is a waste of money, because the general public, which includes the customer, ignores the advertisement as being of no concern to him. More recently, the increase in vandalism and the increase in burglary in peoples' houses and flats increase the probability that everyone will know at least one person who has had his house attacked in this way. Advertising now has a chance of attracting attention, because, rightly or wrongly, it can work on two human characteristics – curiosity about what it must be like to have your house burgled and fear that it might happen to you.

In the context of encouraging awareness of a need, without which implementation cannot start, effective advertisements do three things. They literally show what burglary is like, perhaps with a photograph of a room in chaos after an attack; they have a few words to reach the fear that is latent in us all; and their punch-line concerns what to do about it.

What do we want to see said in that punch-line? The choice inevitably reflects the sponsors of the advertisement. It lies between, first, the police, who have a declared objective of crime prevention and of impartiality, but who have severely limited funds for publicity; second, the security system suppliers, who cannot be expected individually to be impartial; third, the insurance companies; and fourth, the industry's trade associations.

Insurance companies

It is to the insurance companies that I look mainly to generate awareness in the public mind of the need for seeking skilled advice on crime prevention measures and on firms competent to carry out those measures. Individually and collectively they have control of the funds that can be made available for generating awareness, they have a vested interest in reducing losses, and they are mostly independent of the security system suppliers. It is true that insurance people are already a primary source of information, but that is for the enlightened. What we are talking about at the moment is how to get through to the unenlightened.

Trade associations

Their growing strength makes the security industry's trade associations a further source of help to potential and existing customers in need of help and advice.

To round off this part of the problem, it is clearly necessary that whoever is allocated to give advice on security when approached by members of the public should be trained in the subject.

Legislation, standards and inspection

Unlike safety and fire, security has relatively little national legislation to support it and to enforce maintenance of suitable standards. In some countries there is some spin-off from safety requirements, as, for instance, for equipment connected to telephone lines for remote signalling and control. It is important on safety grounds to ensure that no deliberate or accidental connection of mains power supply voltages is made to telephone lines. Not only is there a requirement for type approval of any equipment intended for connection to telephone lines, but also there is a requirement in some countries that any installation of the equipment be accompanied by a contract for maintenance so long as the equipment remains so connected. This virtually rules out direct sale of the equipment, as the only reliable way of ensuring that a maintenance commitment exists is to rent, lease or hire the equipment to the user.

As a further aid, national standards and codes of practice exist to guide companies in good installation and maintenance methods. Standards and codes have little commercial value without the prospect of business sanctions in the event of non-compliance with the standards.

Two of the more recent introductions to 'encourage' compliance are the ACPO Intruder Alarm Policy and the European Commission requirements on electromagnetic compatibility (EMC). These, together with many others, are listed in Chapter 28 and the Appendix.

Various forms of inspectorate exist to present this incentive, operated either nationally or by the industry itself. In the UK, the National Approval Council for Security Systems (NACOSS) has proved increasingly effective in improving implementation of standards. An interesting side-effect on their work derived from the way in which the relevant British Standards were prepared. For instance, I was involved in the preparation of BS4737, covering intruder alarms in buildings. In the main it was drafted by technical people representing member companies of the industry's trade association, the British Security Industries Association. By the time it was published as a British Standard, it had been edited by legal experts, in such a way that it became difficult to interpret in practical terms in the field. NSCIA (National Supervisory Council for Intruder Alarms, as it was then) provided effective help in interpreting the Standard, and with that experience they were a knowledgeable bridge between industry and the British Standards Institution in subsequent revisions of that Standard and in the preparation of others.

As mentioned above, the security industry has little national legislation to support it. The UK is a case in point, but the principal trade association, the BSIA, has grown in stature to become recognised as the self-regulatory body to the industry. It is responsible, with the Association of British Insurers, for the operation of NACOSS, and as part of its self-regulatory function it set up in 1987 the Security Systems Inspectorate and the Manned Services Inspectorate.

System design

Perhaps more than anything else, it is system design that establishes the character of the supplying company. One firm becomes known for being able to deal with a given type of risk in one way, while another firm might approach that risk in another way. Similarly, it is likely to be installation design and implementation, as dealt with in the next chapter, that establishes the reputation of a company.

If system design is to give character to a firm, it is dependent upon the range of security devices available to it. Thus, system design cannot be done in isolation. First, it must take account of the risk analysis, and select detection methods and locations to match the risks; and second, it must keep costs within the contract price. This emphasises the need for the company to have a standard compatible range of sensors, interconnection methods and control and signalling equipment.

Nor can system design be done in isolation from installation design. Life becomes difficult if the risk analysis calls for space detection in the approaches to a risk, and system design selects surface-mounted microwave sensors, which the lady of the house will not have visible at any price.

It is from such situations that the need for a new product can arise.

Establishing a case for a new product

Logical argument has to be part of a designer's stock in trade, and the argument takes the form of a process of elimination – eliminating one idea after another by counter-arguments until you are left with only one idea, or maybe two, which cannot be shot down within the framework of the design objective. I call this the 'clay-pigeon' technique. If an idea stands up against everything that can be fired at it, there may be some merit in the idea and it could be worth following up.

That is only the starting point in establishing a case for a new product. In a small firm the rest of the process may be covered by intuitive thinking by the decision-maker in the firm, but with larger organisations a more formal assessment is called for.

The idea itself may be good enough to get authority to carry out a feasibility study, including the making of an experimental model to show whether it works, or can be made not to work by the villain. Many an idea fails at this point, particularly if the investigation team is well drilled in being villains.

If the idea stands up, and gets approval for further investigation, information has to be assembled on such things as potential sales, acceptable selling price, production costs versus quantities produced, reliability, ease of installation and servicing, availability of people and resources to cover development, manufacture, training and marketing, and the effect on sales of existing products, particularly if the new idea has a chance of becoming a market leader.

There is always someone who will resist new ideas, and properly so if the aim is to preserve a standard range of products which all personnel understand and can handle without excessive supervision. The new product may have to be reserved for the special projects side of a business until it is established, when it can be made available for standard installations. Thus, the specials of today become the standards of tomorrow.

A compelling argument for firm A to add a new product to its range comes if firm B has already adopted it and is taking business away from firm A. If firm A decides that it is better to act on its own initiative rather than react to situations forced upon it, then it will have to accept and gear itself to cope with the problems of being first with a new product.

Organisation for system design

We have seen above two examples of how a possible need for new products can divert the normal activity of system design. A well-organised firm can probably take the consequences in its stride. But if we look at the example of the grille door discussed at the beginning of this chapter, the organisation adopted for incorporating system design can spell success or failure for the firm. The system design for the grille door, as far as it went, was competently done, and yet the intruders were so easily able to defeat the system as installed.

How far, then, must system design go into the detail of each product to ensure that the responsibilities of the system designer are satisfied? How far should it have been necessary to specify the welding of the sheet to the bars to the frame? Should it be necessary to say that pop rivets are not acceptable? There has to be a common-sense limit to the detail of specifications. Beyond that limit the system designer is surely entitled to rely upon the internal supervision of product design and manufacture to meet internal codes of practice and external standards of good security engineering practice, all of which should be based on answers to the dual question, 'Will it work? Can the villain defeat it?'

If it is agreed that the system designer is entitled to rely upon his colleagues to interpret his design in a competent manner, should he take any further interest in the project? If not, the situation looks dangerously like the horizontal structure that has been the downfall of so many companies. If vertical or 'profit centre' organisation is in operation, then the system designer has a right as part of the team to inspect the equipment 'as made', to check that design interpretation is acceptable.

Horizontal organisation may be easier to manage, but that is not the point. The management problems posed by the system designer apparently being able to 'interfere' with the work of another department have to be faced and solved if the customer is to be satisfied. A method adopted by numerous successful firms is to make the systems engineer the project leader or project manager, and to give him authority to see the project through all departments.

Progress towards customer satisfaction has also been made in those countries that have national or regional inspectorates, authorised to carry out independent inspections of installing companies and installations carried out by them. The allocation of the title 'Approved Installer' indicates a standard of competence to be maintained by that company, and it is some guide for the customer new to security in deciding who to approach for advice.

Discussion points

If you have not thought about it before, you will probably be able to recognise the structure, or maybe the lack of it, of your own organisation. If organisation is second nature to you, the problems of successful project management are likely to be fruitful topics for discussion. But if you do come up with something useful, try not to leave it at that. It may be worth going through procedures like those needed for getting an idea for a new product accepted to get your idea on organisation implemented.

How far can you get in drawing distinctions between the functions of risk analysis, system design, installation design and product design? Can you identify the areas of overlap? Can you justify any encroachment by system design into installation and product design?

31 Installation design and implementation

Some practising security installation engineers will probably be tempted to read this chapter first, as being apparently the most relevant to their daily work. They are very welcome to do so if they wish, but I have to say that although this chapter, like others, contains practical hints, tips, ideas and suggestions, it is not the purpose of the book to give detailed installation instructions. What I am aiming at is understanding and awareness in all that is done. The villain is looking for loopholes, and it is no use my telling you to plug this loophole and that loophole, because the villain will find another. If you are involved in installation design, it is up to you, on site, to find the loopholes peculiar to that site, to be aware of what could happen there and to provide a security installation that will alert those concerned that a villain is there.

Principles and practice

A saying that really says it all is: 'Give a man a fish, and you will feed him for a day. Teach a man how to fish, and you will feed him for life.' Detailed installation instructions which will feed you for today are better found in national and company standards and codes of practice. In this book we shall continue our efforts to feed you for life. Hopefully, the efforts may also help the writers of the national and company standards and codes of practice themselves.

In previous chapters we have considered risk analysis and system design, which are concerned with *what* should be done. Installation design deals with *how* it should be done – the decisions on just what equipment should be used where on site.

An alarm company is probably doing well if it gets one order for every three quotations it makes. Put another way, a company probably has to carry out at least three different site surveys for every single order it gets. As the surveys are 'free' or, more specifically, as the cost of the unsuccessful surveys has to be loaded into the price of the successful ones, competitive reality compels most firms to carry out only the risk analysis and system design necessary to enable them to prepare a quotation.

Actual installation design is normally done after an order has been received, because it is itself quite a costly exercise, involving one or more people spending a fair time on site and further time at base finalising the installation information for the field engineers. Although doing the installation design work after an order has been received can save costs, it also presents consequent cost risks, because the detail engineering surveys may reveal the need for more equipment or more time on site than the pre-quotation system surveys revealed.

The smaller the project the more likely is a company to take this risk. With larger projects, where the competition may be less and the customer's risk are higher, it is more likely that the system designs and the installation design work would be done together before a quotation was prepared.

Now for some of the principles of installation design.

Installation design and people

It may be surprising that I tend to put the customer's power supply as a first priority in site work. Historically, burglar alarms consisted only of door contacts and pressure mats, all connected in parallel on normally open circuit, so that no power was consumed until a circuit was closed by, say, someone treading on a pressure mat. Thus, dry cell primary batteries provided all the power needed for the year or more of the shelf-life of the battery. This was fine until the villain found that he could make a door contact useless by cutting the wire to it.

The cure was to connect the devices in series, on normally closed circuit. This meant that power was needed continuously, not just when an alarm signal was required. Closed-circuit operation had to become standard practice to answer the dual question, 'Will it work? Can it be made (by the villain) not to work?' And so rechargeable batteries had to become standard practice. This required mains power for float or recharging of batteries. Thus, the mains power is vital in that it provides the life blood of the system. Cut that and the system dies. The merit of the old open-circuit system is that it dies later.

In some customer premises it is the practice to switch off at the mains for overnight and weekend periods. What is needed is a 'maintained supply', one that is not switched off at any time, and is normally provided to maintain a limited range of vital services in the building. To make sure that an extension from the maintained supply is provided, and provided in the right place, often needs investigation by the works electrician or the outside contractor, and it takes time to get an answer and a supply from them. It is worth making the power supply your first priority on site: otherwise you may find yourself having to wait for other people.

Quotations should always say when the customer and not the alarm company is to provide the maintained mains power supply point.

A feature probably settled at the alarm system design stage is the method of alarm signalling. Where telephone lines are to be used, installation design includes deciding where the line is to be run, or rerun if an existing line is suitable, to limit the risk of it being cut by an intruder.

Already we have mentioned two features that cannot be settled without the help of the customer. As we go on in the installation design sequence, liaison with the customer is necessary again and again, so an important feature is to arrange with the customer, as was done for the system survey, to nominate a person who can act as your opposite number, someone very familiar with the detail of the building and the normal and abnormal working routines of the customer.

While you have his attention it is as well to go through with your opposite number the procedure to be followed in setting the alarm and in locking up and opening up, and the actions involved should an alarm occur, real or false. He will be able to tell you whether certain parts of the building can be locked and alarmed while overtime work continues in another part of the building. If these details were not brought out during the system surveys before quotation, they may affect the detailed design of the control panel envisaged for the project. However, in view of the well-founded emphasis by police on the avoidance of false alarms, it may be desirable or necessary to resist customer attempts to have the building zoned into working and non-working areas unless separate final exit doors can be arranged.

To complete this concept of a final exit door, the last to leave does need an indicator to show him that the system is not in an alarm condition when he is ready to lock up. If there is a risk of his ignoring even that warning, then the risk can be avoided by solenoid interlocking to prevent him physically from locking up until he has cleared the fault internally – for example, by closing a door that may have been left open.

The options on door control are clearly quite numerous, but once the final exit door control details have been cleared with the customer, the question of keyholding has to be faced. It is customary for the police to require names and addresses of two keyholders, who may be members of the customer's staff, of the alarm company, of a guarding company or of the firm running the central alarm station, if employed on the project. It is not the function of the police to break into premises, and a keyholder is required to attend the premises in the event of an alarm to give access to the police to investigate and to the service engineer, who is required to silence any audible alarm still sounding, and to deal with any technical matters arising from the alarm call. Keyholding details need to be settled with the customer before the system is put into service.

Installation design and equipment

After discussing such aspects as these with your escort, it is likely that he will be anxious to get on with something else, and be quite happy to leave you to get on with things that you can do better on your own.

Where perimeter detection devices have been specified, the installation design should be checked to ensure that the siting of the devices and the required sensitivity can give the greatest possible immunity from false alarms without undue risk of failing to detect actual intrusion. In crucial risks, and where there has to be acceptance of a risk of evasion in order to avoid false alarms, then it is customary to back up the system internally with other devices such as space or motion detectors. A system design may have omitted these, the designer not having realised just how critically otherwise the perimeter sensors would have to be set up.

More often than not, space detection is specified in its own right, rather than as a back-up to other detection devices. The reasons come from operational system design, where it is considered that roof penetration and movement from one part of the building to another cannot be detected in any other way, and from competitive system design, which may show that space detection is cheaper in equipment plus labour costs than alternative forms of perimeter detection.

Installation design comes in in checking that the most suitable form of space detection has been chosen. In the main, the choice lies between ultrasonic, microwave and passive infrared sensors. Each has its virtues and characteristics which make it more or less suitable for a particular environment than others. These characteristics are explained and illustrated in Chapter 15, 16 and 17, and every surveyor and installation design engineer needs to know them. This means that every firm needs to allow its people time to 'play' with these devices on the lines suggested in those chapters, so that any lingering myths are blown away by personal experience and understanding. A common check that has to be made is that microwave sensors, which can 'see' through glass, are sited not to look at windows, where legitimate passers-by could be seen by the sensor and a false alarm caused.

The typical alternative is ultrasonics, but here the detection range is often shorter, and as the wavelength also is shorter, false alarms from vibration, including vibration of window-panes from traffic and aircraft, still have to be considered when the type of sensor to be used and its location are decided upon. As described in the Part 2, a location often overlooked, and one that is particularly suitable for microwave sensors, is overhead, using ceiling or roof-truss mounting. Owing to directional sensitivity, windows can be made 'out of sight', and a longer overall detection range can often be achieved with relatively low detection sensitivity settings.

A point regarding the use of passive infrared sensors is raised in the suggestions for discussion.

Organisation of field work

So, by taking a few examples some of the principles involved in installation design have been illustrated. We could have gone on to consider cable routes and so on, but even for the newcomer the treatment should have served to show that installation design is a function separate from surveys and system design. As regards the field engineer it will, hopefully, encourage him to understand why things are done as they are and to be alert to things that could go wrong.

As regards the manager, time is taken up with so many distractions that he rarely takes a long hard look at how his field work is organised. If there is one subject that he could profitably focus his attention upon, it is, 'What are the things that hinder and stop field engineers from getting on with the job?'

Have you been on site lately yourself, for instance, and worked through the general and specific information given to the field engineer on that site to find out whether the information is clear and sufficient for uninterrupted work? Have you checked how the kits of equipment are made up in stores for the field engineers? Do the kits include any special tools needed? Or should the installation design engineer have been instructed to look out for the need for extra-long ladders or masonry drills so that stores can provide them when needed?

Have satisfactory arrangements been made for the field engineers to have access to the site during their normal working hours? Are your skilled field engineers wasting their time running many hundreds of metres of cable high overhead in a warehouse, when this could be better done by an acceptable electrical subcontractor? The security of the system is unlikely to be prejudiced provided that the subcontractors never connect the wiring to the equipment.

Can incentives be effective in encouraging field engineers to complete on time without doing inferior work? Does field supervision, inspection and test have a bias towards overseeing the field engineers or towards customer satisfaction? The distinction can play a significant role in field personnel morale and attitudes.

If customer satisfaction is the objective, there is much to be said for having a supervisor carry out the final stages of implementation, including user training, preparation and explanation of the completion report for signing by the customer, and for hand-over of keys. It may be more expensive in present costs, but the customer is made to feel more important this way; it helps to train field engineers in customer liaison; and it leads to better feedback of overall field information to management.

Discussion points

Listening to a group of installation engineers recalling the problems that can arise in the field is like listening to a series of isolated incidents and to them that is perhaps what they are. But listen again and see whether there is a characteristic thread running through them. If you can identify just one thread, be it concerned with paperwork, access to premises, the performance of equipment, the opinion of customers or traffic conditions, it is worth setting up informal discussions from which needed improvements can be identified.

Some surveyors tend to select a type of sensor on hunch rather than on detailed knowledge of its suitability for a particular location. Earlier in this chapter attention was drawn to the need for installation designers to verify that the most suitable sensor is installed for each location. That is not the same thing as deciding to use, say, infrared just because both microwave and ultrasonic sensors might be troublesome in a given location. A discussion between surveyors, installation designers and equipment designers should focus on the circumstances in which detection devices are selected on merit and the circumstances in which they are selected because alternatives have been excluded for one reason or another. Given good leadership, a rather mature discussion should develop which would be a helpful training session and also would provide guidance material for future instructions to the various categories of field personnel.

32 User training, operational orders and monitoring

You have done your best. The risk analysis was good and you unearthed some problems that the customer had not thought of when he first enquired about the provision of an intruder alarm system at his premises. The system design took account of the risks adequately without making the overall cost of the project uncompetitive, and both installation design and implementation had drawn compliments from the customer for clean, tidy and unobtrusive workmanship.

And yet you feel uneasy at the thought of what will happen to the system when you and your team have left the site and left its operation in the hands of the user. It's not so bad if the user is the individual allocated as your opposite number by management during the planning stages, because you will have had opportunities to talk to him and to brief him on what to expect. The unease reappears when you realise that he cannot do everything himself and that he will have assistance from others, such as keyholders and the like, whom you may never have met.

Establishing the need for training

The unease is at its worst on sites manned round the clock by uniformed security personnel. Not only have you not met them all because of shift working, but also the chances are that you do not speak their language. What chance is there, then, of all the users learning and having the interest and confidence to operate the system correctly? Unless you can persuade management that they have wasted the money invested in the security system if they do not also invest in adequate training of their personnel who are to use the system, the chance is negligible.

A production manager would not dream of installing an expensive new type of machine tool without having his staff trained to use it. So, to win over the production manager, you can talk of the security system as being a tool, an aid to cost reduction through loss reduction. At all levels of management, similar parallels can be found to help you speak the managers' language, and so convince them of the need for training.

In this chapter, and elsewhere in the book, the user is variously described as the customer, management, the escort or opposite number representing the customer during system design, the keyholder, the chief security officer, the uniformed security officer on patrol or on duty in a security central control room, and the civil or military police officer on duty in his own type of security office. Each, directly or indirectly, is concerned with calling for the attention of reaction forces to deal physically with intruders in the event of an alert signal being raised by the security system.

Clearly, the particular category of user depends upon the size and nature of the customer's organisation, and the nature of the training programme needs to be slanted to match the type of user. Up to a point, the training can be carried out by a member of the firm supplying the security system but, as indicated above, the job is better done by someone who can express the ideas in words that can be understood by his audience. So, the task of the system supplier in these circumstances is to train the trainer.

Training material and methods

What, then, are the principles involved in preparing and presenting training material? Start, as always, with the objective: what are we training these people to do? The answer may be arrived at if we consider the typical daily life of operational security people, who may spend most of the time on routine security and non-security matters. Then, suddenly, they have to act. The objective is that they should act with understanding, to pass on to the reaction forces without delay all predetermined and exceptional information needed for the reaction forces to carry out their work.

A key principle in teaching them understanding is to teach them 'what to expect'. The obstacles to quick effective action are fear and uncertainty. Those obstacles are least likely to survive if the user has been taken through, in words and actions, the situations that are associated with normal working, full alert, false alarms and also the unexpected. Words and actions are emphasised, for in situations where quick response is necessary only personal practical experience is enough to let each individual feel he really does know what to expect.

If you are still in doubt about the need for practical experience, it is worth remembering the old adage, 'No impression without expression'. Its validity is arguable, but, to take not too extreme an example, it seems easy to tell someone how to use a spanner to remove a nut, say, from a bicycle wheel. However, if you watch anyone do it for the first time, even after being told how, he will show doubt and uncertainty as to whether to turn the spanner clockwise or anticlockwise. Even if he has a 'feel' or aptitude for that sort of thing, it will do no harm in his training if you make him meet the unexpected – a left-hand thread, in which the spanner has to be turned in the opposite direction to normal.

Attitudes

Another key principle in training is that operational people need to be able – and willing. Once you have given them the ability to use a security system, why ever should they be unwilling to use it?

The most potent argument in selling a security system to operational personnel is that it is an aid to them, to help them in their job of keeping villains away. If, however, the number of security personnel was reduced when the security system was introduced, you have a problem. The remaining personnel are only human if they think that their jobs, too, could go if they made too great a success of the new security system. It is unusual for organisations to be overstaffed with security personnel, and security systems are added to deal with a worsening security situation as an alternative to increasing manpower.

So, in a well-planned system, personnel problems arising from job losses should be minimal, but it is necessary to be aware of potential reluctance to make electronic aids effective. Probably the quickest route to disillusion with a security system is via false alarms, and a corresponding proportion of training time and attention needs to be devoted to explaining them. Rather than making site security personnel passive victims of false alarms, they can be a crucial source of information, and it helps all round if they can be involved in the job of tracing these troubles by being asked to log specific information, together with anything they think relevant when false alarms occur.

A more subtle source of unwillingness to make a system work satisfactorily is management attitude. Security personnel are quick to notice if management ignore security instructions given to staff in general, and in designing a training programme due care needs to be given to persuading management of their responsibility to set a good example in observing security standards.

Management, too, need to be included in briefing and training on the significance of real alarms and the actions that would be taken. They will sort out their own procedures, but they need to be reminded of the need to be ready for an emergency arising from a loss or intrusion: Who is to assess and check the loss? Who is to deal with the insurance company? Who is to liaise with the police? And if the press become interested, who is to be the spokesman?

Operational orders

In Chapter 29 a section was devoted to the theme of speaking the user's language. This is of particular significance in preparing operational orders, and a chief security officer's comment to me, 'Tell me what you want my men to do, and I will write an order in words that they will understand', was quoted.

Speaking the user's language is equally significant in training, and the language to be used in training may be different even within the range of users involved. Who is the best man to act as 'interpreter'? Fortunately, it seems, as indicated above, the chief security officer is the most likely to be the man, since very often he has had military service training and experience and has added to that the skills in security needed for him to secure the post as chief security officer. He, then, is the natural focal point for training, operational orders and liaison with management.

Supervision

But it is not enough to train people and to give them operational orders. How are we to know that the orders have been carried out, and correctly? It is often said among security people, 'If we cannot trust our security colleagues, who can we trust?' Just think what it means if you reply 'No one' to that question. You, I and our best friends can slip up, and it is no disgrace to be supervised. All levels of management are failing in their duty if they do not supervise those they are responsible for. It is the way supervision is done that matters. If the attitude in supervision is to help, as one would help a friend, rather than 'to catch him out', then we can have an atmosphere in which personnel are likely to co-operate with any electronic monitoring or automatic supervisory equipment that is provided, rather than resist and oppose it.

Just as in aircraft, automatic recording equipment can be provided to monitor 'alert' situations, their time of origin, their location and actions taken, and, just as important, they can confirm nil reports when nothing happened. Comparison of recorded data with operational logs provides a fair monitor of how the electronic security system meets its responsibility of giving warning of attack, and of how security manpower meets its responsibility of acting as instructed and trained to do.

On unmanned sites it would be significant to have the locking up and opening times recorded, together with the time and location of any incident causing an alarm signal.

The value of automatic monitoring equipment to operational personnel, to management and to system service engineers is such that it needs to be discussed during the planning stage and included, when practicable, in the overall system concept proposals. The installed location of the equipment also needs consideration, and the need for impartiality may well lead to its being installed away from the operational security control equipment. As suggested above, the attitude to automatic monitoring equipment by senior personnel will have an important bearing upon its acceptability among operational personnel.

One key to a successful attitude may be found in an installation where closed-circuit television was proposed for monitoring shop-floor personnel on high-risk work. Resistance to the use of CCTV in this instance ceased when it was agreed that the equivalent of the shop steward could also have a TV monitor showing the same information that the CCTV system would provide to management – an inexpensive solution to what could have been quite a problem.

Discussion points

In one brief chapter we have touched upon three potentially vast subjects. But the emphasis is not so much on the subjects themselves as on the interplay of the subjects with one another in the security environment. It emphasises the uselessness of security systems without people, and without people who understand what they should do, and who do it.

Each aspect of the links between training and operational duty, and between operational duty and supervision, raises its own series of questions, but probably the one most worthy of consideration is, 'What is needed to establish and to maintain the attitudes required for co-operation by all concerned in the use of security equipment towards its objective of crime prevention?'

Too often this type of discussion can lead to a session full of complaints that gets nothing done. This section of the book deals with Implementation, so, at whatever level of operation you find yourself, act on any proper conclusion that is within your own power to do.

33 Maintenance and service

When I told a friend that I wanted a car, he said, 'Oh, you should buy a so and so, they provide such a splendid spares service'. 'But I want', I replied, 'I want a car that does not need a spares service like that'. I was looking for reliability, and it seemed to me that a car that needed a splendid spares service was likely to be unpredictably unreliable.

Similar ideas can apply to most manufactured products, but so far as security systems are concerned, there is one major difference. When you buy a car, you buy a fully factory-assembled single piece of equipment, with responsibility for reliability centred on the factory. With the majority of security systems several factory-made pieces of equipment are brought to site and installed, and interconnecting wiring is run to complete the overall assembly of the system. The site work is not carried out by the people who made the equipment, and probably it is carried out by people controlled by totally different employers. Thus, responsibility for the reliability of the overall system is split – split between the equipment manufacturer and the system installer.

So, what are we to regard as a proper attitude to reliability and to the consequences of unreliability – that is, to the need for maintenance and service?

Designing for reliability

First, let us try to get a scale of values, a yardstick by which to judge the significance of reliability. A car might be run 24 000 miles in a year at an average speed of 40 miles per hour, giving a total running time of 600 hours a year. An intruder detection system in lock-up-and-leave premises might be in use from 6 p.m. to 8 a.m. for five nights a week plus day and night for Saturday and Sunday, giving 118 hours a week, or, when multiplied by 52 weeks, 6136 hours a year, compared with 600 hours a year for the car.

Had you realised that a typical security alarm system has to run reliably for at least ten times as long each year as a typical car?

When we set out to design a car or security equipment, the objective has to match the customers – do we want to sell at the lowest possible price, or do we want to sell the maximum possible reliability? The customer faces the decision as to whether to pay a higher first cost to get reliability or pay a lower first cost, with higher maintenance and service costs, to compensate for lower reliability.

If we are to design for reliability, where does the money go that has to be recovered in higher first costs? Practically all intruder detection devices, control and signalling equipment incorporate semiconductor diodes, transistors and integrated circuits, which are inherently reliable in principle and are manufactured in such vast quantities that the high development and production engineering costs needed to achieve sustained reliability in use can be spread over the quantities sold. The cost is there, but the cost per unit is small.

Simple tolerancing

If, however, we insisted upon every single device of any given type being exactly the same, the cost would be enormous and unsellable. This argument is easier to follow, perhaps, in the case of discrete components such as resistors and capacitors, which are given their rated value, plus or minus an allowance, or tolerance, in value either side of the rated value.

If all components were to be made exactly to the values required by the designer, the excessive cost would be balanced to some extent by lower design costs, because the designer's job would be relatively easy.

Accepting tolerances on the values of the components means that their unit cost is lower, but the designer is involved in much more work in ensuring that the equipment will operate correctly, no matter what combination of tolerances of components happen to come together during manufacturing assembly. Not only that, but, should a component fail after a few years, the equipment must operate satisfactorily again when the component is replaced by one of the same rated value, but perhaps with its tolerated deviation from the rated value being at the opposite extreme. The designer is therefore involved in designing not only for initial manufacture, but also for future maintainability and serviceability. And the costs of this work have to be reflected in the price.

Total tolerancing

Taking account of the effect of variations in the value of a few components, simple tolerancing, was manageable, but for total tolerancing the sheer scale of calculating the possible combinations of tolerances in a complete piece of electronic equipment and their effect upon performance was so extensive and expensive that it was rarely done, with consequent contributions to unreliability and unserviceability.

All that changed, or could change, with the advent of computer-aided design. The effects of combinations of tolerance, the selection of appropriate tolerance values, and the design changes needed to allow for the tolerance combinations can all be calculated and assessed so much more quickly. Thus, further substantial improvements in electronic equipment reliability and serviceability are feasible. If these opportunities are taken, the equipment manufacturer is near to meeting his responsibility for reliability.

Near – but not near enough. With resistors the principal rating is the number of watts the resistor can dissipate as heat without the temperature of the resistor becoming excessive. Reliability does not depend only on the designer deciding on the correct wattage to use when his circuit is working normally; he must also consider, with or without the aid of a computer, whether that resistor would be damaged by overheating if another component in the circuit failed. This aspect of design is called consequential damage. With capacitors the principal rating is the applied voltage that the capacitor can withstand without damage or destruction.

And remember: the manufacturer of the individual components is himself in a competitive world, and the safety factors in the rating of components made by one manufacturer may be less than those made by another manufacturer. Thus, the concept of total tolerancing has to include the safety factors built into the components themselves, lest all the good work done on equipment tolerancing be wasted in the name of material cost reductions.

In spite of the complexity of problems facing the designer, of which the above examples are but a few, and of the hazards facing the manufacturer, the overwhelming evidence is that manufacturers are meeting their responsibilities and that electronic equipment as such is remarkably reliable.

The fact that security systems as installed are relatively unreliable must mean that the unreliability originates outside the individual electronic assemblies.

Connectors

The next step in a search for the reasons for system unreliability is to look at installation work. The normal interface between the electronic designer, his equipment and the installer is the wiring connector, be it a plug and socket, barrier strip, terminal block, tab connector, wire wrap, crimped sleeve or whatever was selected by the equipment designer. It is at this interface that so many installation and service problems arise, as one or two examples from life will show. Once you are alerted by these examples to the interconnection problem, some of the reasons for it and possible solutions, you should be better able to look at equipment and assess its suitability for installation, servicing and use.

In Chapter 9 the curious properties of copper were highlighted. So long as copper remains the normal material for interconnecting wire, these properties need to be understood. As supplied in wires and cables, the copper is 'half-hard' – that is, it is neither too soft nor too hard. In this condition it is strong and easy to bend. In bending wire, however, the inner portion is compressed and the outer portion is stretched, and the curious effect of bending is to make the copper brittle, to 'work-harden' it. Bend the wire a few more times and it will break.

Now relate this problem back to the equipment interconnection device. How many of you have struggled to get wire through the casing of the equipment and pass it along the narrow gap between the side of the casing and the terminal block, and then, having bent the end of the wire at right angles, tried to get the end into an invisible hole in the terminal block? And failed? You try again, straighten the wire, bend it again, and success at last. Not really, because if the terminal block has pinch-screws for securing the wire in the block, screwing down the pinch-screw on to the copper wire compresses the copper over a small concentrated area and makes it brittle. Provided that the wire is not disturbed from then on, the connection will probably survive reliably, but in routine maintenance, and particularly in fault-finding, service engineers sometimes need to disconnect parts of the system. If disconnection is made at the terminal block just mentioned, the wiring will be disturbed, with the consequent risk of wire fracture, not only of the wire which the service engineer is disconnecting and which he can repair, but also of an adjacent wire in the block which has been work-hardened and embrittled by the pinch-screw. Service engineers have a name for the law which says that that fracture will not show up until some time after they have left the site.

There is no need to put up with this sort of thing. It is up to the designer to recognise copper for what it is, to design into the equipment those types of connecting block that do not compress copper unduly, and to locate the blocks in the equipment casing to allow the installation and service engineer to see what he is doing and to get the wires into position with the minimum of bending.

I had just such a problem recently. A consulting engineer not only suggests to a client that something should be done this way and not that way, but also is often required to demonstrate that it can be done, and sometimes is commissioned to carry out the final project. In this instance all requirements were met in the very small space available by using a terminal block with wire entry and wire clamping screw slots at 45° to the base, to give visibility and accessibility, while the excessive wire compression pressure was avoided in the block by the use of load distribution bars between the screw and the wire. Ample pressure for electrical contact is achieved in this way, without undue compression and work-hardening of the copper wire, and connections were made first time, with minimum bending.

Some connecting blocks use the right principle but the wrong application of the principle, in using *thin* springy strips between the screw and the wire. These have only limited success in pressure distribution, and score mainly in avoiding tearing stranded wires apart as a result of screw rotation, as distinct from pressure. They can fail, for instance, when a wire is removed and some roughness catches on the strip and pulls it out of the hole. It is a tantalising job getting the strip back in the hole. The connectors can succeed, however, if manufactured with the ends of the strips simply formed as an inverted loop. But the connectors are by no means exclusive, and the choice can lie with any connector that takes account of the properties of copper and of what can happen on site.

Installing for reliability

Interconnection cables

The choice of interconnection cables has a bearing on site reliability. Awareness of the properties of copper, as described above, gives the clue here. The thinner the strand of wire the less the work-hardening when the wire is bent but the lower its mechanical strength and current-carrying capacity. The thicker a strand of wire the greater the work-hardening and the more difficult the wire to bend and manage. Unfortunately, nature is against us in that for the optimum gauge or thickness of wire for electronic interconnections we seem to get the worst combination of work-hardening risk and ease of handling. That optimum would give us a single strand of wire, but single-strand wires have a bad history of failure through fracture in service. The simple solution is to use multiple strands of thin wire twisted together to form a multistrand cable. Although still prone to work-hardening, this is much more reliable than single-strand cable.

Electrical interference

The next range of problems are likely to be environmental, in the form of various types of electrical interference, conducted, induced or radiated.

Although legislation covering EMC came into force in 1996 outlawing the generation of, and susceptibility of equipment to, interference (beyond defined limits), it is important to be aware of the three main types of interference as they could still be the cause of site problems in spite of the legislation. The most effective way of avoiding site problems is to ensure that the equipment is designed with the appropriate suppression built in.

Conducted interference   This is mostly generated by the various forms of high-speed electronic switching of power equipment connected to the electricity power supply mains. When supply mains are used to maintain battery supplies for security equipment, mains-borne interference can find its way into the security equipment and cause erratic and unwanted switching of various of its functions.

There is an argument that such interference should be suppressed at source instead of being allowed to pollute the mains supply. In those countries where no legislation exists to control this form of interference the normal procedure is to suppress it retrospectively at the premises suffering the interference. There is also an argument for building suppression into the security equipment at the design and manufacturing stage. The decision turns on the certainty of increased initial cost with the possibility that the precaution is unnecessary in a fair number of system installations. Another certainty is that the cost of retrospective action on site for those installations where suppression proves necessary is far higher than the cost of building in suppression at the manufacturing stage.

Induced interference   This tends to be continuous, and to be at a level too low to cause unwanted switching yet high enough to interfere with wanted signals. Again it normally originates from mains power supplies, from mains cables lying too close to signal cables causing hum on audio equipment or hum bars on CCTV monitor screens.

Radiated interference   The worst form of radiated interference comes from lightning, and, owing to the very high energy involved, there is also induced interference. Statistically there is a high incidence of false alarms coinciding with these electric storms, and for short distances away from the storm centres there seems little that can be done about it. For the inner fringes outwards, however, the methods adopted for limiting the effects of switching spike voltages on power supply mains inputs to equipment can be made to assist in reduction of interference from electric storms.

Another source of radiated interference is, perhaps obviously, from radio transmitters. If radio masts are nearby, they give ample warning of their potential for trouble, but not so readily traced are fleeting vehicle-borne transmitters such as CB radios and police radios.

Perhaps the least obvious but, fortunately, well-known source of continuous interference is the gas discharge lamp – or fluorescent light. Some not so well-designed microwave motion detectors can be held in the alarm condition when installed close to a lighted gas discharge tube.

Radio and gas discharge interference normally responds to treatment with low-inductance (flat, not rolled) capacitors connected close to the component most susceptible to interference and ground. If the interference is being picked up by the component itself, as in an inductor, for instance, the component needs to be screened with a metallic cover connected to ground.

Earthing

In semiconductor electronics there is rather more carelessness or unawareness regarding grounding or earthing than is justified. Owing to the very low values of current involved and the short lengths of leads or printed circuit track, only negligible voltage drop develops along the lead, and it does not, in general, matter too much where along the track various components are connected to ground.

Where interference signals are involved, the current and, hence, the voltage drop in earth leads can be substantially higher than normal equipment working levels. To avoid malfunctioning from interfering signals, it helps if as many as practicable of the leads to ground are all taken to a single point and if one side of the power supply is connected to that point only.

Another sometimes neglected practice is the physical connection of one or other side of the power supply to ground, via the mains supply earth, or a water pipe, or a ground stake. Although it is not always successful, I have found that physical grounding is more conducive to long-term trouble-free operation of permanently installed equipment than allowing the system to 'float' with respect to ground.

A particularly tiresome fault to trace can be in systems where considerable care has been taken to use the single-point earthing principle. That earthing point is usually at the power supply, and any screening used to protect cables from interference is also connected to that point. To maintain single-point earthing, the screen has to be sheathed in an insulating sleeve, and after the system has been in operation satisfactorily for some time, it is possible, especially in outdoor installations, for the insulating sleeve to deteriorate or wear away owing to abrasion in wind. Once the screen is exposed, it can make metallic contact with something local and already earthed such as a fence, and an earth loop is formed which can upset the effect of screening and the single-point earthing. If the outdoor earthing is intermittent, the problem of tracing the fault is that much more difficult.

Battery life

In Chapter 27, we discussed the use of batteries to maintain security systems in operation for a predictable period of time after a mains failure or series of failures. Almost regardless of the choice of battery used, sooner or later the battery will fail to hold its charge for the predicted and required duration of time. Unless included in a routine schedule of maintenance tests, it is all too easy to overlook this long-term effect; and even in a well-ordered maintenance programme, it is time-consuming to conduct a convincing test.

Such a test would be to have the security system 'on', but to switch off the system battery charger at a given time. From then on, at regular intervals, readings of battery voltage and, if possible, load current are taken until the system goes into 'fail safe' alarm owing to insufficient battery supply voltage. Alternatively, the test can be concluded when the battery voltage is down to, say, 1.8 V per cell.

Ideally, the battery voltage is then plotted against time, and the graph compared with the results of previous tests on the system. If the battery only just lasts the required time, then clearly it will not be long before it falls below specification, and it should be replaced.

The disconcerting point is reached, perhaps if the battery has not been tested before, when the system goes into alarm in the first half-hour of the test where, say, it was expected to go on for at least 8 h. With lead–acid batteries this often follows from float charging at too high a voltage and from neglect to top up the battery with distilled water: and the battery has become 'dry', with no chance of holding its charge. For wet lead–acid batteries a float charge voltage of 2.2 V per cell can limit gassing and loss of electrolyte for normally acceptable periods. For sealed recombination-type lead–acid cells the voltage can be somewhat higher, as discussed in Chapter 27.

Human problems

Moving on now from the purely physical problems in maintenance to problems involving human nature, one of the hardest to design out of a system is user misuse, and consequently we tend to be resigned to live with it. There is no malicious intent in user misuse, just what I choose to call 'ignor-ance' of quite reasonable operating instructions, hamfistedness in forcing control knobs beyond their natural end-stops, and 'redecorating' infrared sensor housings, including painting over the filter glass.

Daytime damage is another matter. One cold store suffered again and again from cables being cut during working hours; apparently, our system interfered with a brisk behind-the-counter trade. So many problems such as this go unnoticed until it is time to set the alarm at night, with the result that service departments become inundated with urgent service demands all at much the same time.

Quite a lot can be done to avoid the '6 p.m. fault rush-hour' by making better use of the tamper circuits commonly incorporated in the wiring but switched off with the remainder of the security system during normal working hours. How much better it would be to leave the tamper circuit alive during working hours, with an audio-visual tamper alarm installed in the chief security officer's office or elsewhere, where it can be acted upon immediately. There would then be a much better chance of a service engineer being available to you well before closing time.

Another human problem concerns keyholding. If for any reason the service engineer has to call at lock-up-and-leave premises after closing time, he has a problem in gaining access. Neither the police nor the insurance company are likely to provide a keyholding service, so the owner has to appoint one or more of his staff as keyholders to attend the premises on call-out with the service engineer. This is by no means a popular appointment, and the logical step in all but strictly security terms is to appoint the security company itself as keyholder. Except for the higher risks, it may be considered adequate to issue the keys to the service engineer attending the premises; for major risks a more acceptable arrangement is for the service engineer to be accompanied by a uniformed security officer as keyholder.

Visual maintenance

Actual faults apart, the more one considers servicing the more one seems to be tempted towards the philosophy of leaving well alone. This certainly does not mean doing nothing, but, as we saw earlier in this chapter and elsewhere, electronic equipment itself is remarkably reliable, and so is interconnection cabling, provided that it is not disturbed too much. Leaving equipment and cabling alone needs to be complemented by regular and thorough visual maintenance.

One of my earliest 'service jobs' was as a schoolboy, when I was called in by a neighbour whose radio had stopped working. By pure luck (or instinct, if you will) I found the trouble in, I suppose, something under a minute. My neighbour had an outdoor aerial, popular at the time, with the lead-in stapled round the skirting board to the receiver. Looking closer, I noticed a green patch on the lead-in wire; the wall was obviously damp and verdigris corrosion had severed the wire. That corrosion provides an example of visual maintenance, in that it could have been noticed and cured long before complete failure had occurred.

Back with security systems: a host of simulated tests can be carried out simply by representing a villain in a non-destructive way to trigger various of the sensors incorporated in the system. This has the particular advantage of including the environment in the tests and being as true to life as possible.

When the control panel is to be checked, it is important to get the user to go through his routines. You know what should happen, and if the user rather than you operates the system, you are more likely to spot misuse and causes of potential trouble.

Probably in preparing their routine maintenance and call-out service methods alarm service companies will combine the best features as they see them of both physical maintenance and 'leave well alone' techniques. All I urge is that they keep their methods under regular review in the light of their own and others' experience, not only to keep down costs, but also to enhance the reputation of security systems for reliability, and progressively to get false alarms down to a level at which the deterrent effect of security systems is not overshadowed in the eyes of the police by a waste of their effort involved in attending false calls.

Maintenance contracts

If maintenance is so important, it is worth considering how to get it built into the various forms of system supply contracts. To an extent the method of payment will be dictated by the purchasing company's financial policy and resources. Beyond that there is choice.

The most positive way of having maintenance included is entering into the equivalent of full rental contracts. As the name implies, the customer never buys the equipment itself, paying only for the use of it, and, whether he realises it or not, for near-full depreciation or amortisation of the equipment within the rental period. The rental charge is at a fixed rate for part or the whole of the rental period, so it is in the interests of the supplier to maintain the equipment well in order to keep down his variable costs arising from call-out service.

An intermediate arrangement between full rental and full purchase exists whereby equipment connected to telephone lines is rented, together with a maintenance commitment. The remainder of the equipment in the installation may, however, be purchased by the customer, with some latitude in choice as to whether the equipment bought is also covered by a maintenance contract.

For installations which are entirely 'in-house', signalling only to the purchaser's own security personnel on site, and also for small shop-type installations having only a local alarm bell and no connection to telephone lines for remote signalling, the choice ranges from outright purchase of the whole system with no maintenance commitment, through to full rental again.

Within that range there is an arrangement that finds favour with the larger organisations, for systems that are essentially reliable. This option allows for occasional routine maintenance calls, say every 6 months, but for no call-out service calls under the contract. Any call-out service visit is on a one-off basis, and instead of the charge being on an averaged basis for regular plus call-out visits the charge for call-outs is separate, covering travelling, subsistence, time and materials used on that particular visit.

Resurveys

In the long term a factor of interest to the insurance company as well as to the user and the alarm company is the resurvey concept. As businesses evolve, some additions or alterations to premises occur, with consequent changes of use. These changes tend to accumulate unnoticed, so if the alarm company offers a resurvey of the premises every 5 years, say, there is an opportunity to take care of obsolescence, change of use, additions and deletions, and, above all perhaps, in selling the idea to the customer the alarm company can avoid the risk of the insurance company declining to pay out a claim on the grounds that the security system no longer matched the risk. Dealing with that problem is maintenance in its fullest sense.

Discussion points

The full story of routine maintenance and call-out service can never be told, but if you are really interested in the subject, do try to get among two or three service engineers and encourage them to talk shop with you. There is endless fascination in the problems they meet and the tricks they get up to in order to solve them. Where better, too, for a designer to learn what things need to be improved?

34 Feedback

As you start to read this chapter, can you trace back in your mind how the book came to be open at this page? If the book just happened to open here, and you started to read out of curiosity, there would be little or no evidence of control in what happened, up to that point. In order to read, however, a complex series of biological functions take place which commonsense tells us just could not lead to intelligible results unless there was some sort of control on those functions. And that control could not work effectively without some knowledge of the objective. If this sounds over-complex, let us try something simpler.

Anyone who has had to deal with a domestic water storage tank is likely to be familiar with the ball valve used for the control of water supply to the tank. The objective is clear. As water is drawn from the tank for domestic use, we require the water used to be replaced until the water level in the tank regains the predetermined level, automatically.

The device normally fitted to the tank to control the flow of water consists in essence of a float, coupled by an arm to a flap, which covers the end of the water inlet pipe. As water is drawn from the tank, the float follows the water level down, and in so doing the arm removes the flap from the end of the water inlet pipe, water flows in and the water level rises. The float rises with it until it reaches its original position, at which point the flap again covers the inlet pipe and water flow into the tank ceases. The flap *can* be designed so that there is no flow of information between it and the float until the flap is closed suddenly when the required water level is reached. If that happens, the water in the supply pipe is likely to object to being stopped, and it may force the flap open again, to let some more water in, and so on.

The right thing to do, surely, is to design the flap so that there is a constant flow of information between it and the float and back again regarding their relative positions, so that the rate of water flow is reduced as the float nears its 'full' position and a sudden stop to water flow is avoided. This would then be a simple form of information feedback control to achieve a given objective automatically.

I tend to think of that as a static situation, and we are more likely to meet a dynamic situation in, say, linear electronics. If we have an amplifier with an irregular frequency – amplitude characteristic, we can take a sample of the output and feed it back to the input so that the net input is reduced somewhat – negative feedback. If the output is higher than is required at any given frequency, then the sample fed back will be higher and the net input reduced, leading, if the process is quick enough, to the objective – a linear response.

If the objective is to find a given page in a book, the mind is pre-programmed or trained to understand the nature and sequence of numbers. The book is opened at the wrong page. The mind calculates how many pages back or forward it must instruct the hand and fingers to turn to correct the error. The error may then be only one page, and very delicate control or touch of the fingers may be needed to select and turn just one page to reach the objective. Again this is an example of feedback of information from the eye to the mind to the finger to the eye to the mind, and so on, until the errors are corrected.

In human terms, we can have anarchy, meaning general lawlessness and disorder, or civilisation, where concern for others leads to an accepted way of life and to methods of control to correct deviations from that way of life. Our role in security is in the detection of deviations and in feeding back information to those charged with the duty of correcting deviations. We are, therefore, very much part of the feedback loop needed to maintain stability.

The name 'cybernetics' has been given to the science of control by feedback of information in material and biological systems. It is interesting that, in both material and human terms, overcontrol leads to unexpected side-effects, undercontrol leads to objectives not being achieved and just good tight control leads to discipline.

This book, too, is part of the feedback loop referred to. Whatever experience I may have gained is set down to help new and practising security people to avoid errors and practices that have already been found to be counterproductive.

To that extent it is negative feedback, and the book can be used for reference to ways of doing things that can be successful. But what really appeals to me is that feedback can also be positive. A major objective of the book would be achieved if controversy made it a catalyst for the generation of new ideas. So much of what we have today didn't even exist ten years ago and creative thinking needs to be stimulated to help the measures for crime prevention to take control of the situation.

However, be aware that positive feedback can be explosive and control of it has to be expected. Control cannot be effective unless the objectives are known, and objectives cannot be defined unless the needs of the user are known. Isn't that where we came in?

British Standards for security equipment and systems

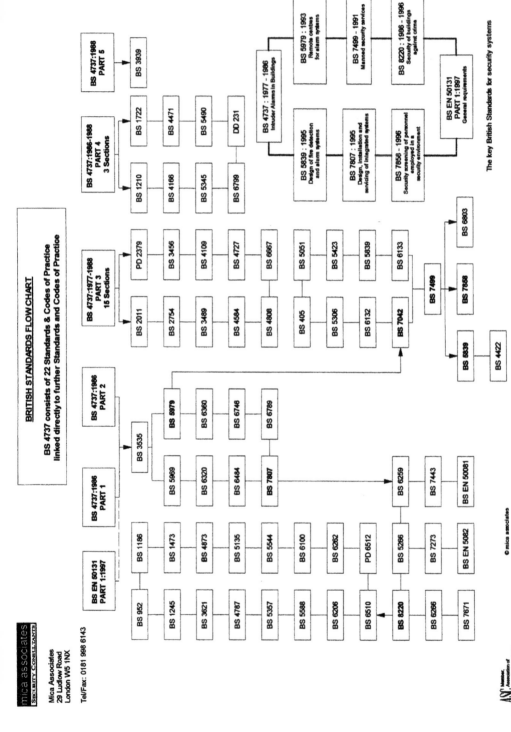

BRITISH STANDARDS FLOW CHART

BS 4737 consists of 22 Standards & Codes of Practice
linked directly to further Standards and Codes of Practice

mica associates
SECURITY CONSULTANTS

Mica Associates
29 Ludlow Road
London W5 1NX

Tel/Fax: 0181 998 6143

The key British Standards for security systems

© mica associates

ASC National Association of Security Consultants

Index

Definitions of terms are indicated in **bold** type.

Printed and bound by CPI Group (UK) Ltd, Croydon, CR0 4YY

03/10/2024

01040430-0012